D0773035

Surviving the SOC Revolution

A Guide to Platform-Based Design

Henry Chang
Larry Cooke
Merrill Hunt
Grant Martin
Andrew McNelly
Lee Todd

KLUWER ACADEMIC PUBLISHERS
Boston / Dordrecht / London

Contents

Distributors for North, Central and South America:
Kluwer Academic Publishers
101 Philip Drive, Assinippi Park
Norwell, Massachusetts 02061 USA
Telephone (781) 871-6600
Fax (781) 871-6528
E-Mail <kluwer@wkap.com>

Distributors for all other countries:
Kluwer Academic Publishers Group
Distribution Centre
Post Office Box 322
3300 AH Dordrecht, THE NETHERLANDS
Telephone 31 78 6392 392
Fax 31 78 6546 474
E-Mail <orderdept@wkap.nl>

Electronic Services <http://www.wkap.nl>

Library of Congress Cataloging-in-Publication Data

Surviving the SOC revolution: a guide to platform-based design/Henry Chang...[et al.].
  p. cm.
  1. Embedded computer systems--Design and construction. 2.
Computers--Circuits--Design and construction. 3. Application specific integrated
circuits--Design and construction. 4. System design. I Chang, Henry.

TK7895.E42 S87 1999
004.16--dc21   99-052726

Printed on acid-free paper.

Printed in the United States of America

Authors

Henry Chang, Ph.D. is a senior member of consulting staff in Cadence Design Systems Design Methodology Engineering group, and is the chair of the VSI Alliance's mixed-signal development working group.

Larry Cooke is an independent consultant to the EDA and electronics design industry.

Merrill Hunt is a fellow of Cadence Design Systems.

Grant Martin is a senior architect in Cadence Design Systems Design Methodology Engineering group.

Andrew J. McNelly was a senior director of Solutions Architecture in Cadence Design Systems Strategic Marketing group, and is currently the senior director of Strategic Marketing at Simutech, LLC.

Lee Todd was a senior director of Solutions Development in Cadence Design Systems Strategic Marketing group, and is currently the senior director of Business Development and Product Marketing at Simutech, LLC.

Acknowledgments

A collective work of this nature is not possible without a considerable amount of mutual support. For that, we would like to thank and acknowledge each other's contribution to the book. The support and encouragement of many other people were also essential for achieving our goal. We would like to acknowledge Steve Glaser, Bob Hon, Patrick Scaglia, Joe Mastroianni, and Alberto Sangiovanni-Vincentelli for their key support. The pivotal roles of Doug Fairbairn, Diana Anderson, and Larry Rosenberg in establishing and growing the VSI Alliance deserve special acknowledgment. Shannon Johnston, Fumiyasu Hirose, Eric Marcadé, and Alain Rabaeijs reviewed the material and provided very useful comments and suggestions. A special debt of gratitude is owed to Linda Fogel, our editor, whose relentless pursuit of consistency, clarity, and completeness was critical to pulling the diverse styles and thoughts of all the authors together. This book might never have been finished without her tireless contributions.

We would also like to thank Maria Pavlick, Mary Stewart, Gloria Kreitman and Cathereene Huynh for their hard work in getting this volume to production.

Cadence Design Systems provided the fertile environment for discussing and exploring the ideas in this book. As the book took shape, the corporate marketing team was instrumental in seeing to the details of layout, graphics, cover design, and handoff to the publisher.

Many of the ideas in this book were built after extensive discussions with many different people. In particular, we would like to thank Mike Meyer, Jim Rowson, Rich Owen, Mark Scheitrum, Kent Shimasaki, Sam George, Pat Sheridan, Kurt Jagler, Leif Rosqvist, Jan Rabaey, Dan Jefferies, Bill Salefski, Stan Krolikoski, Paolo Giusto, Sanjay Chakravarty, Christopher Lennard, Jin Shyr, Kumar Venkatramani, Pete Paterson, Tony Kim, Steve Manser, Graham Matthew, and Ted Vucurevich for their part in these discussions. The particular conclusions drawn in this book, of course, are the responsibility of the authors alone. There are many others not specifically named who also contributed, and to them we would like to extend our thanks. In addition, each author would like to especially acknowledge the following people.

Henry Chang would like to pay special thanks to his wife, Pora Park, and to his colleagues on the VSI Alliance's mixed-signal development working group.

Larry Cooke would like to acknowledge with heartfelt thanks his wife, Diane, and their son, David, for their unwavering support during the long hours of writing.

Merrill Hunt would like to acknowledge his wife, Pamela, and children for their support and encouragement. He would also like to acknowledge his appreciation of the many chip designers with whom he has worked and shared successes and failures.

Grant Martin would like to pay special thanks to his wife, Margaret Steele, for constant support and encouragement, and to his daughters, Jennifer and Fiona, for their willingness to tolerate his moods as the deadlines approached.

Andrew McNelly would like to especially acknowledge the unending support and encouragement of his wife, Merridee, and son, Trent.

Lee Todd would like to thank his wife, Linda, for her support and guidance as the process for pulling the book together wore on in the final stages. He would also like to thank the systems designers with whom he has worked, who shaped many of the perspectives in this book.

Henry Chang
Larry Cooke
Merrill R. Hunt
Grant Martin
Andrew J. McNelly
Lee Todd

San Jose, California
July 1999

Preface

By the year 2002, it is estimated that more information appliances will be sold to consumers than PCs (*Business Week*, March 1999). This new market includes small, mobile, and ergonomic devices that provide information, entertainment, and communications capabilities to consumer electronics, industrial automation, retail automation, and medical markets. These devices require complex electronic design and system integration, delivered in the short time frames of consumer electronics. The system design challenge of the next decades is the dramatic expansion of this spectrum of diversity. Small, low-power, *embedded* devices will accelerate as microelectronic mechanical system (MEMS) technology becomes available. Microscopic devices, powered by ambient energy in their environment, will be able to sense numerous fields, position, velocity, and acceleration, and communicate with substantial bandwidth in the near area. Larger, more powerful systems within the infrastructure will be driven by the continued improvements in storage density, memory density, processing capability, and system-area interconnects as single board systems are eclipsed by complete systems on a chip.

The overall goal of electronic embedded system design is to balance production costs with development time and cost in view of performance and functionality considerations. Production cost depends mainly on the hardware components of the product. Therefore, to minimize production cost, we must do one of the following:

- Tailor the hardware architecture to the functionality of the product so that the minimum cost solution is chosen for that particular application, or
- Determine a common denominator that could be shared across multiple applications to increase production volume.

The choice of one policy over the other depends on the cost of the components and on the agreements on costs versus volume in place with the manufacturers of the hardware components (IC manufacturers in primis). It is also rather obvious that the common denominator choice tends to minimize development costs as well. The overall trend in industry is in fact to try to use a common hardware "platform" for a fairly large set of functionalities.

As the complexity of the products under design increases, the development efforts increase exponentially. To keep these efforts in check, a design methodology that favors reuse and early error detection is essential.

Both reuse and early error detection imply that the design activity must be defined rigorously, so that all phases are clearly identified and appropriate checks are enforced. To be effective, a design methodology that addresses complex systems has to start at high levels of abstraction. In most of the embedded system design companies, designers are familiar with working at levels of abstraction that are too close to implementation so that sharing design components and verifying designs before prototypes are built is nearly impossible.

Design reuse is most effective in reducing cost and development time when the components to be shared are close to the final implementation. On the other hand, it is not always possible or desirable to share designs at this level, since minimal variations in specification can result in different, albeit similar, implementations. However, moving higher in abstraction can eliminate the differences among designs, so that the higher level of abstraction can be shared and only a minimal amount of work needs to be carried out to achieve final implementation.

The ultimate goal is to create a library of functions and of hardware and software implementations that can be used for all new designs. It is important to have a multilevel library, since it is often the case that the lower levels that are closer to the physical implementation change because of the advances in technology, while the higher levels tend to be stable across product versions.

We believe that it is most likely that the preferred approaches to the implementation of complex embedded systems will include the following aspects:

- Design costs and time are likely to dominate the decision-making process for system designers. Therefore, design reuse in all its shapes and forms will be of paramount importance. Flexibility is essential to be able to map an ever-growing functionality onto an ever-evolving hardware.
- Designs have to be captured at the highest level of abstraction to be able to exploit all the degrees of freedom that are available. Such a level of abstraction should not make any distinction between hardware and software, since such a distinction is the consequence of a design decision.
- Next-generation systems will use a few highly complex (Moore's Law Limited) part-types, but many more energy-power-cost-efficient, medium-complexity ((10M-100M) gates in 50nm technology) chips, working concurrently to implement solutions to complex sensing, computing, and signaling/actuating problems.
- Such chips will most likely be developed as an instance of a particular platform. That is, rather than being assembled from a collection of

independently developed blocks of silicon functionality, they will be derived from a specific "family" of micro-architectures, possibly oriented toward a particular class of problems, that can be modified (extended or reduced) by the system developer. These platforms will be extended mostly through the use of large blocks of functionality (for example, in the form of co-processors), but they will also likely support extensibility in the memory/communication architecture as well.

- These platforms will be highly programmable.
- Both system and software reuse impose a design methodology that has to leverage existing implementations available at all levels of abstraction. This implies that pre-existing components should be assembled with little or no effort.

This book deals with the basic principles of a design methodology that addresses the concerns expressed above. The platform concept is carried throughout the book as a unifying theme to reuse. This is the first book that deals with the platform-based approach to the design of embedded systems and is a stepping stone for anyone who is interested in the real issues facing the design of complex systems-on-chip.

Alberto Sangiovanni-Vincentelli
Chief Technical Advisor
Cadence Design Systems, Inc.

Rome
June 1999

1

Moving to System-on-Chip Design

The continuous progress in silicon process technology developments has fueled incredible opportunities and products in the electronics marketplace. Most recently, it has enabled unprecedented performance and functionality at a price that is now attractive to consumers. The explosive growth in silicon capacity and the consumer use of electronic products has pressured the design technology communities to quickly harness its potential. Although silicon process technology continues to evolve at an accelerated pace, design reuse and design automation technology are now seen as the major technical barriers to progress, and this productivity gap is increasing rapidly. As shown in Table 1.1, the combination of increasing complexity, first and derivative design cycle reductions, design reuse, and application convergence creates a fundamental and unprecedented discontinuity in electronics design. These forecast levels of integrated circuit (IC) process technology will enable fully integrating complex systems on single chips, but only if design methodologies can keep pace.

Incremental changes to current methodologies for IC design are inadequate for enabling the full potential for system on chip (SOC) integration that is offered by advanced IC process technology. A paradigm shift comparable to the advent of cell library–driven application-specific integrated circuit (ASIC) design in the early 1980s is needed to move to the next design productivity level. Such a methodology shift needs to reduce development time and effort, increase predictability, and reduce the risk involved in complex SOC design and manufacturing.

The required shift for SOC design rests on two industrial trends: the development of application-oriented IC integration platforms for rapid design of SOC devices and derivatives, and the wide availability of reusable virtual components.

The methodology discussed in this book is based on the evolution of design methodology research carried out over many years. This research was first applied to physical IC design, then refined for constraint-driven, analog/mixed-signal

Table 1.1. Evolution of Silicon Process Technology

	1997	1998	1999	2002
Process technology	0.35µ	0.25µ	0.18µ	0.13µ
Cost of fab	$1.5 - 2.0 billion	$2.0 - 3.0 billion	$3.0 - 4.0 billion	$4.0 billion+
Design cycle	18 - 12 months	12 - 10 months	10 - 8 months	8 - 6 months
Derivative cycle	8 - 6 months	6 - 4 months	4 - 2 months	3 - 2 months
Silicon complexity	200 - 500K gates	1 - 2M gates	4 - 6M gates	10 - 25M gates
Applications	Cellular, PDAs, DVD	Set-top boxes, wireless PDA	Internet appliances, anything portable	Ubiquitous computing, intelligent, interconnected controllers
Primary IP sources	Intragroup	Intergroup	Intercompany	Intercompany, interindustry

(AMS) design, broadened to deal with hardware/software co-design for reactive systems, and finally, generalized in system-level scope to deal with the full range of embedded SOC design problems. The methodology has immediate applicability, as well as the range and depth to allow further elaboration and improvement, thus ensuring its application to SOC design problems for many years to come.

The interest in consumer products in areas such as communications, multimedia, and automotive is the key economic driver for the electronics revolution. The design of embedded consumer electronics is rapidly changing. Changes in the marketplace are demanding commensurate changes in design methodologies and toolsets. Some of the market-driven forces for change are:

- Shrinking product design schedules and life spans
- Conforming products to complex interoperability standards, either de jure (type approval in communications markets) or de facto (cable companies acceptance of the set-top market)
- Lack of time for product iterations due to implementation errors: a failure to hit market windows equals product death
- Converging communications and computing into single products and chipsets

These forces have a profound effect on design methodology. This book addresses what needs to be done to bring design technology in line with IC process technology. It explores ways to look at the problem in a new way to make this transition as quickly and painlessly as possible. To better understand the direction we need to go, we need to examine where we stand now in the evolution of design methodology.

The Evolution of Design Methodology

The transition from transistor-based to gate-based design ushered in ASIC, provided huge productivity growth, and made concepts such as gate arrays a reality. It also fostered the restructuring of engineering organizations, gave birth to new industries, and altered the relationship between designer and design by introducing a new level of abstraction.

Historically, our industry seems to follow a cycle: IC process technology changes, and design technology responds to the change with creative but incomplete solutions. Design methodology then adapts these solutions to the new process, creating incremental increases in productivity. During the more dramatic periods, such as the one we are currently in, a major leap up the abstraction curve is needed to exploit the process technology. With that leap, the industry undergoes a fundamental reorganization—design is not just done faster, it is done differently by different people, and it is supported by different structures. Over the past 25 years, this has occurred about every 10 years with a three-year overlap of driving methodologies.

We are now entering the era of block-based design (BBD), heading toward virtual component-based SOC design, which is driven by our ability to harness reusable virtual components (VC), a form of intellectual property (IP), and deliver it on interconnect-dominated deep submicron (DSM) devices. In just a few years, the silicon substrate will look like the printed circuit board (PCB) world as shown in Figure 1.1, and reusable designs will be created and packaged as predictable, preverified VCs with plug-and-play standard interfaces.

What Is SOC Design?

To begin, we need to define SOC design in a standard and industrially acceptable way. The Virtual Socket Interface (VSI) Alliance, formed in 1996 to foster the development and recognition of standards for designing and integrating reusable blocks of IP, defines system chip as a "highly integrated device. It is also known as system on silicon, system-on-a-chip, system-LSI, system-ASIC, and as a system-level integration (SLI) device."[1] Dataquest has defined an SLI device as having "greater than 100 thousand gates with at least one programmable core and on-chip memory."[2]

1. VSI Alliance *Glossary*, VSI Alliance, March 1998.
2. ibid.

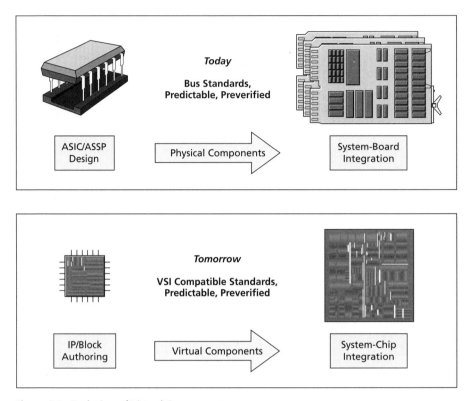

Figure 1.1. Evolution of Virtual Components

In this book, SOC design is defined as a complex IC that integrates the major functional elements of a complete end-product into a single chip or chipset. In general, SOC design incorporates a programmable processor, on-chip memory, and accelerating function units implemented in hardware. It also interfaces to peripheral devices and/or the real world. SOC designs encompass both hardware and software components. Because SOC designs can interface to the real world, they often incorporate analog components, and can, in the future, also include opto/microelectronic mechanical system (O/MEMS) components.

The Electronic Industries Association of Japan (EIAJ) has defined an Electronic Design Automation (EDA) Technology Roadmap for designing a "cyber-giga-chip" by the year 2002.[3] This design incorporates DRAM, flash memory, CPU cores, digital signal processor (DSP) cores, signal processing and protocol control hardware, analog blocks, dedicated hardware units, and on-

3. "Cyber-Giga-Chip in 2002," EDA Technofair handout, EIAJ EDA Technology Roadmap Group, February 1998.

chip buses. This is a good illustration of the complexity that future SOC designs will need to achieve.

Linchpin Technologies

Discontinuities caused by a change in silicon process technology demand that new design technology be invented. Linchpin technologies are the building blocks for transitioning to the next level of design methodology. Typically, the new technology is partnered with an ad hoc methodology adopted early on to form systems *effective enough* at addressing the first set of design challenges to deliver products. Because these new technologies offer significant improvements in design capability, functionality, and cost, as well as creating a change in design methodology and engineering procedures, they are recognized as essential steps for broad change to occur.

Looking back on the evolution of design technology, many linchpins are easily identifiable (see Figure 1.2). For example, gate-level simulation enabled an increase in design verification capacity sufficient to address the silicon capacity potential. But designing within the bounds of the gate-level logic meant accepting modeling accuracy limitations of the simulator and associated libraries, which resulted in a fundamental design methodology change. Similarly, register-transfer level (RTL) synthesis technology facilitated an increase in designer productivity, but required the transition to RTL-based design capture, and verification and acceptance of the predictability limitations of optimization technology. Often the linchpin technologies are cumulative, that is, they are built upon each other to make a synergistic improvement in

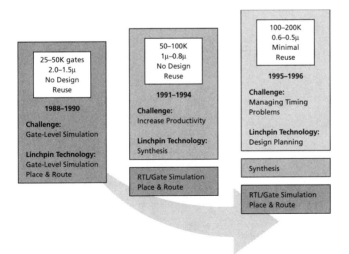

Figure 1.2. Historical Linchpin Technologies

productivity. They also must support the mix of legacy designs that use previous design methods.

Design Methodologies

The primary design methods used today can be divided, as illustrated in Figure 1.3, into three segments: timing-driven design (TDD), BBD, and platform-based design (PBD). These segments vary depending on the linchpin technologies used, the design capacity, and the level of and investment in design reuse.

Looking at the electronic design market in this way helps to identify where a given design team is in the design methodology evolution. It also helps in determining which design technologies and methodologies are needed to facilitate the transition to the next step. History has shown that the companies that can make the transitions the fastest have success in the market.

Note, however, that there are gray areas between segments where some design groups can be found. Also, the transition process is serial in nature. Moving from TDD to PBD is a multistep process. While larger investments and sharper focus can reduce the total transition time, a BBD experiential foundation is necessary to transition to PBD.

The following sections describe the design methodology segments, and identify the necessary linchpin technologies and methodology transitions. Table 1.2 summarizes some of the design characteristics that pertain to the different methodologies.

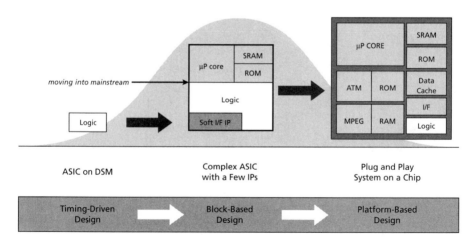

Figure 1.3. Primary Design Methodologies

Table 1.2. Summary of Design Characteristics

Design Characteristics	TDD	BBD	PBD
Design complexity	5000 to 250K gates	150K to 1.5M gates	300K gates and greater
Design level	RTL	Behavioral/RTL	Architecture and VC evaluation
Design team	Small, focused	Multidisciplinary	Multigroup, multidisciplinary
Primary design	Custom logic	Blocks in context, custom interfaces	Interfacing to system and bus
Design reuse	None	Opportunistic soft, firm, and hard	Planned firm and hard
Primary optimization focus	Synthesis, gate-level architecture	Floor planning, block architecture	Silicon-compatible system architecture
Primary design granularity	Gates and memory	Functional clusters, cores	VCs
Bus architecture	None/custom	Custom	Standardized/multiple application-specific
Test architecture	None/scan	Scan/JTAG/BIST/custom	Hierarchical, parallel scan/JTAG/BIST/custom
Mixed-signal	None	A/D, PLLs	Functions, interfaces
Constraint/goal specification	Logic constraints	Block-budgeted constraints	Interface constraints
Verification level	RTL/gate	Bus functional to cycle accurate/RTL/gate	Mixed (ISS to RTL with hardware and software)
Hardware/software co-verification	None	Hardware/software functionality and interfaces	Hardware/software interface only
Partitioning focus	Synthesis limitations (hierarchical)	Functions (hierarchical)	Function/ communications (hierarchical)
Placement	Flat	Hierarchical	Hierarchical
Routing	Flat	Flat	Hierarchical
Timing analysis	Flat	Flat with limited hierarchy	Hierarchical
Delay calculation	Flat	Flat	Hierarchical
Physical verification	Flat	Flat with limited hierarchy	Hierarchical

Timing-Driven Design

TDD is the best approach for designing moderately sized and complex ASICs, consisting primarily of new logic (little if any reuse) on DSM processes, without a significant utilization of hierarchical design. The design methodology prior to TDD is area-driven design (ADD). In ADD, logic minimization is key. Design teams using this methodology tend to be small and homogeneous. When they encounter problems in meeting performance or power constraints that require shifting to a TDD methodology, the following symptoms are often observed:

- Looping between synthesis and placement without convergence on area and timing
- Long turnaround times for each loop to the ASIC vendor
- Unanticipated chip-size growth late in the design process
- Repeated area, power, and timing reoptimizations
- Late creation of adequate manufacturing test vectors

These symptoms are often caused by the following:

- Ineffective or no floor planning at the RTL or gate level
- No process for managing and incrementally incorporating late RTL design changes into the physical design
- Pushing the technology limits beyond what a traditional netlist handoff can support
- Ineffective modeling of the chip infrastructure (clock, test, power) during floor planning
- Mishandling of datapath logic

DSM technology exacerbates the interconnect management weaknesses of the wire load-based delay model. The inaccuracies of this statistical model become severe with DSM and lead to non-convergence of the constraint/delay calculations. Today's product complexity, combined with radically higher gate counts and shorter time to market (TTM), demands that design and verification be accelerated, trade-offs be made at higher levels of design, and interconnect be managed throughout the process and not left until the end. These all argue against ADD's synthesis-centric flat design approach.

A more floor plan-centric design methodology that supports incremental change can alleviate these problems. Floor planning and timing analysis tools can be used to determine where in the design the placement-sensitive areas are located. The methodology then allows placement results to be tightly coupled into the design optimization process.

Going from RTL to silicon represents the greatest schedule risk for designs that are timing-, area-, or power-constraint driven. Typically, this is managed by starting the physical design well before the RTL verification is completed.

Overlapping these processes reduces TTM and controls the schedule risk, but at the expense of higher non-recurring engineering costs. To successfully execute concurrent design in a flat chip environment requires change-management and floor-plan control processes that are able to incorporate the inevitable "last bug fixes" in the RTL into the physical design and still keep the optimizations already accomplished. Table 1.3 summarizes the benefits and challenges of TDD.

Table 1.3. TDD Benefits and Challenges

TDD Benefits	TDD Challenges
Netlist handoff is well understood; floor plan-centric method limits iterations due to timing discrepancies for most designs.	High-performance, high-density DSM products still require iterative loops from place and route back into timing analysis/synthesis.
Flat-timing analysis/delay calculation is more accurate and better supported by most of today's tools.	Design changes resulting from RTL simulation are very difficult to incorporate in the physical design, impacting TTM.
Accommodates RAM/ROM blocks as cells. Compilers can be used to improve designer productivity.	Flat methodology begins to fail as complexity increases and gate counts rise above 150K.
Test generation can be uniform and automatic.	Synthesis tools limit block sizes; wire-load models are insufficient for high-performance or low-power designs.
Can handle small, soft VC blocks as design sources.	Turnaround time through ASIC physical design is long and subject to non-convergence.

TDD Linchpin Technologies
TDD relies upon the following linchpin technologies:

* *Interactive floor-planning tools* These give accurate delay and area estimates earlier in the design process, thereby addressing the timing and area convergence problem between synthesis and place and route.

* *Static-timing analysis tools* These enable a designer to identify timing problems quickly and perform timing optimization across the entire ASIC. The designer can perform most functional verification at RTL with simpler timing views, reduce the amount of slower timing-accurate gate-level simulations, and rely upon static timing analysis to catch any potential timing-related errors, thereby improving productivity significantly.

- *Using compilers to move design to higher abstractions with timing predictability*
  For example, a behavioral synthesis tool can be linked to a datapath
  compiler, providing an operational vehicle for planning and implementing
  datapath-dominated designs rapidly. This moves critical decision trade-offs
  into the behavioral level, while backing it up with a high-performance
  path to an efficient datapath layout. Applied appropriately, it can radically
  improve a design's overall performance. It also introduces layout
  optimization at the system level, which is needed in block- and
  platform-based designs.

Block-Based Design

Increasing design complexity, a new relationship between system, RTL, and
physical design, and an increasing opportunistic reuse of system-level functions
are reasons to move beyond TDD methodology. Symptoms to look for in
determining whether a BBD methodology is more appropriate include:

- The design team is becoming more application-specific, and subsystems,
  such as embedded processing, digital data compression, and error
  correction, are required.
- Multiple design teams are formed to work on specific parts of the design.
- ASIC engineers are having difficulty developing realistic and
  comprehensive testbenches.
- Interface timing errors between subsystems are increasing dramatically.
- The design team is looking for VCs outside of their group to help
  accelerate product development.

Ideally, BBD is behaviorally modeled at the system level, where
hardware/software trade-offs and functional hardware/software co-verification
using software simulation and/or hardware emulation is performed. The new
design components are then partitioned and mapped onto specified functional
RTL blocks, which are then designed to budgeted timing, power, and area constraints. This is in contrast to the TDD approach, where the RTL is captured
along synthesis-restriction boundaries. Within limited application spaces (highly
algorithmic), behavioral synthesis can be coupled with datapath compilation
to implement some of the new functions.

Typically, many of the opportunistically reused functions in BBD are poorly
characterized, subject to modification, and require re-verification. The programmable processor cores (DSPs, microcontrollers, microprocessors) are
imported as either predictable, preverified hard or firm (netlist and floor plan)
blocks, or as an RTL design to be modified and re-verified. The functional verification process is supported by extracting testbench data from the system-

level simulation. This represents a shift from "ASIC-out" verification to "system-in" verification.[4] This system-in approach becomes the only way to ensure that realistic testbenches that cover the majority of worst-case, complex environmental scenarios are used.

Designs of this complexity usually employ a bus architecture, either processor-determined or custom. A predominately flat manufacturing test architecture is used. Full and partial scan, mux-based, and built-in-self-test (BIST) are all possible, depending on the coverage, design for manufacturability, and area/cost issues. Timing analysis is done both in a hierarchical and flat context. Top-down planning creates individual block budgets to allow synthesis to analyze timing hierarchically. Designers then select either flat or hierarchical extraction of the final routing, with flat or hierarchical detailed timing analysis dependent upon the specific accuracy needs of the design. The design requirements determine the degree of accuracy tolerance or guard band that is required for design convergence. This guard band management becomes especially critical in DSM design. Typical technologies are 0.35μ and below, well within the DSM interconnect effects domain. Design sizes range from 150K to 1.5M gates. For designs below 150K, the hierarchical overhead is not justified, and as designs approach 1.5M gates, PBD's reuse economies are essential to be more competitive.

BBD needs an effective block-level floor planner that can quickly estimate RTL block sizes. Creating viable budgets for all the blocks and their interconnect is essential to achieving convergence. This convergence can be significantly improved through the use of synthesis tools that comprehend physical design ramifications. The physical block is hierarchical down through placement, and routing is often done flat except for hard cores, such as memories, small mixed-signal blocks, and possibly processors.

In BBD, handoff between the design team and an ASIC vendor often occurs at a lower level than in TDD. A fully placed netlist is normal, with many design teams choosing to take the design all the way to GDSII using vendor libraries, hard VCs, and memories, as appropriate. While RTL handoff is attractive, experience shows that such handoffs really only work in BBD when it is supported by a joint design process between the product and ASIC vendor design teams. Without preverified, pre-characterized blocks as the dominant design content, RTL handoff is impractical for all but the least aggressive designs.

---

4. Glenn Abood, "System Chip Verification: Moving From 'ASIC-out' to 'System-In' Methodologies," *Electronic Design,* November 3, 1997, pp. 206–207.

Table 1.4. BBD Benefits and Challenges

BBD Benefits	BBD Challenges
Top-down design approach manages complexity for large designs.	Process does not create a set of reusable components.
Import of external processor cores supported.	TTM constrained by design and re-verification of soft RTL blocks. Process encourages design tweaking of blocks.
Complex, parallel block design teams supported.	Limited availability of cycle-accurate and behavioral models. Tracking models against RTL difficult and expensive.
DSM interconnect effects managed through RTL floor planning (including placement and budgets), datapath, and global interconnect.	Flat-routing and flat-delay calculation approach moves potential interconnect problems to late in the design process.
Link between system and RTL simulations minimizes unique testbench development, and provides confidence that realistic and worst-case scenarios are being verified.	Hardware/software verification often requires a high-performance hardware emulation environment, a consequence of the lack of high-speed, high-level models.

BBD Linchpin Technologies

BBD relies upon the following linchpin technologies:

• *Application-specific, high-level system algorithmic analysis tools* These tools provide productivity in modeling system algorithms and system operating environment. They can be linked to hardware description language (HDL) verification tools (Verilog, VHDL) through co-simulation technologies and standards, such as the Open Model Interface (OMI), and to HDL-based design capabilities, such as RTL synthesis via HDL generation and behavioral synthesis.

• *Block floor planning* This facilitates interconnect management decision-making based upon RTL estimations for improved TTM through faster area, timing, and power convergence. It provides the specific constraint budgets in the context of the top-level chip interconnect. It also supports the infrastructure models for clock, test, and bus architectures, which is the basis for true hierarchical block-based timing abstraction. The ability to abstract an accurate, loaded timing view of the block enables the designer and tools to focus on block-to-block interface design and optimization, which is a key step in reducing design complexity through higher abstraction.

- *Integrated synthesis and physical design* This technology enables a designer to manage the increased influence of physical design effects during the synthesis process by eliminating the need to iterate between separate synthesis and placement and routing tools to achieve design convergence. Using the integrated combination, the synthesis process better meets the top-level constraints in a more predictable manner.

Platform-Based Design

PBD is the next step in the technology evolution. PBD encompasses the cumulative capabilities of the TDD and BBD technologies, plus extensive design reuse and design hierarchy. PBD can decrease the overall TTM for first products, and expand the opportunities and speed of delivering derivative products. Symptoms to look for in determining whether a PBD methodology is more appropriate include:

- A significant number of functional designs are repeated within and across groups, yet little reuse is occurring between projects, and what does occur is at RTL.
- New convergence markets cannot be engaged with existing expertise and resources.
- Functional design bugs are causing multiple design iterations and/or re-spins.
- The competition is getting to market first and getting derivative products out faster.
- Project post-mortems have shown that architectural trade-offs (hardware/software, VC selections) have been suboptimal. Changes in the derivative products are abandoned because of the risk of introducing errors.
- ICs are spending too much time on the test equipment during production, thus raising overall costs.
- Pre-existing VCs must be constantly redesigned.

   Like BBD, PBD is a hierarchical design methodology that starts at the system level. Where PBD differs from BBD is that it achieves its high productivity through extensive, planned design reuse. Productivity is increased by using predictable, preverified blocks that have standardized interfaces. The better planned the design reuse, the less changes are made to the functional blocks. PBD methodology separates design into two areas of focus: block authoring and system-chip integration.

   Block authoring primarily uses a methodology suited to the block type (TDD, BBD, AMS, Generator), but the block is created so that it interfaces easily with multiple target designs. To be effective, two new design concepts must be established: interface standardization and virtual system design.

In interface standardization, many different design teams, both internal and external to the company, can do block authoring, as long as they are all using the same interface specifications and design methodology guidelines. These interface standards can be product- or application-specific.

For anticipating the target system design, the block author must establish the system constraints necessary for block design. Virtual system design creates the context for answering such questions as:

- What power profile is needed?
- Should I supply the block with multiple manufacturing test options?
- Should this be a hard, firm, or soft block, or all three?
- Should there be multiple block configurations and aspect ratios?
- Should this block be structured for interfaces with a single bus or multiple bus types?
- What sort of flexibility should I allow for clocking schemes and internal clock distribution?

System-chip integration focuses on designing and verifying the system architecture and the interfaces between the blocks. The deliverables from the block author to the system integrator are standardized (most likely VSI or internal VSI-based variant) and multilevel, representing the design from system through physical abstraction levels.

Integration starts with partitioning the system around the pre-existing block-level functions and identifying the new or differentiating functions needed. This partitioning is done at the system level, along with performance analysis, hardware/software design trade-offs, and functional verification.

Typically, PBD is either a derivative design with added functionality, or a convergence design where previously separate functions are integrated. Therefore, the pre-existing blocks can be accurately estimated and the design variability limited to the interface architecture and the new blocks. The verification testbench is driven from the system level with system environment-based stimulus.

PBD focuses around a standardized bus architecture or architectures, and gains its productivity by minimizing the amount of custom interface design or modification per block. The manufacturing test design is incorporated into the standard interfaces to support each block's specific test methodology. This allows for a hierarchical, heterogeneous test architecture, supporting BIST, scan-BIST, full and partial scan, mux-based, and Joint Test Action Group (JTAG)/boundary scan methods that can be run in parallel and can make use of the programmable core(s) as test controllers.

Testing these large block-oriented chips in a cost-effective amount of time is a critical consideration at the system-design level, since tester time is getting very expensive. Merely satisfying a test coverage target is not sufficient. Timing analysis is primarily hierarchical and based upon pre-characterized block timing. Delay calculation, where most tight timing and critical paths are contained within the blocks, is also hierarchical. A significant difference from BBD is that the routing is hierarchical where interblock routing is to and from area-based block connectors, and it is constraint-driven to ensure that signal integrity and interface timing requirements are met.

The physical design assembly is a key stage in the design, since most PBD devices are built upon 0.25μ and smaller process technologies, with DSM interconnect-dominated delays. Addressing the DSM effects in the physical interface design is a challenge.

PBD uses predictable, preverified blocks of primarily firm or hard forms. These blocks can be archived in a soft form for process migration, but integrated in a firm or hard form. In some cases, the detailed physical view of the hard VC is merged in by the manufacturer for security and complexity reduction. Firm VCs are used to represent aspect ratio flexible portions of the design. The hard VCs are used for the chip's highly optimized functions, but have more constrained placement requirements. Some block authors provide multiple aspect ratio options of their firm and hard VCs to ease the puzzle-fitting challenge. Less critical interface and support functions can be represented in soft forms and "flowed" into the design during the integration process.

The range of handoffs between designer and silicon vendor broadens for PBD. PBD is likely to begin using the Customer-Owned Tooling (COT)-based placed netlist/GDSII handoff as in BBD. However, as the design becomes dominated by predictable, preverified reusable blocks, a variant of

Table 1.5. PBD Benefits and Challenges

PBD Benefits	PBD Challenges
Planned design reuse yields very high productivity.	Planned reuse requires significant up-front design planning and accurate future product plans.
Designs can be composed of diverse, specialized functions from multiple sources.	Significant software portions require extensive hardware/software co-verification.
Hierarchical routing and timing reduce design focus completely.	Embedded AMS design content creates noise and test issues.
Interface-based design promotes higher abstraction design and implementation.	Platform migration to new process technology requires re-characterization of VCs (hard and soft) and platform architecture.
Multiple reuse for blocks allows for development cost amortization and more optimal block design.	Requires organizational change to support separated block authoring and system-chip integration.

RTL sign-off becomes a viable option. We expect that silicon vendors will provide processes for handing off designs at the RTL/block levels. Successful design factories will depend on such processes both to manage the vagaries of DSM and to meet their customers TTM challenges. In six-month product cycles, physical design cycles of three to five months are not acceptable.

PBD Linchpin Technologies
PBD is enabled by the following linchpin technologies:

- *High-level, system-level algorithmic and architectural design tools and hardware/software co-design technologies* These tools, which are beginning to emerge[5] and will serve as the next-generation functional design cockpit, provide the environment to select the components, partition the hardware and software, set the interface and new block constraints, and perform functional verification by leveraging high-speed comprehensive models of VCs.

- *Physical layout tools focused on bus planning and block integration* These tools, used early in the design process through tape-out, are critical to PBD. The physical design effects influence chip topology and design architecture. Early feedback on the effects is critical to project success. For bus-dominated interblock routing, shape-based routing technology is important. Such tools enable the predictable, constraint-driven, hierarchical place and route necessary for PBD.

- *VC-authoring functional verification tools* As the focus of PBD verification shifts to interfaces—block to block, block to bus, hardware to software, digital to analog, and chip to environment—tools for authoring VCs must evolve to provide a thorough verification of the block function and a separation of VC interfaces from core function. OMI-compliant simulation tools allow co-simulation at various levels of abstraction, from system algorithm/architecture level to gate level. This also enables the environment-driven, system-level verification test suites to be used throughout the verification levels. Emerging coverage tools allow the VC developer to assess and provide VC verification to the integrator.

Reuse—The Key to SOC Design

The previous section discussed the design methodology transitions that a company can go through on the path toward effective system chip design. Now

---

5. G. Martin and B. Salefski, "Methodology and Technology for Design of Communications and Multimedia Products via System-Level IP Integration," *Proceedings of Design Automation and Test in Europe Designer Track,* February 1998, pp. 11–18.

we'll look at the same transition path from a VC and design reuse perspective. Note that the methodology and reuse evolutions are mutually enabling, but do not necessarily occur in unison.

As we move forward in the transition to SOC, TTM assumes a dominant role in product planning and development cycles. The National Technology Roadmap for Semiconductors asserts that design sharing is *paramount* to realizing their projections.[6] Reuse is a requirement for leadership in the near term and survival in the medium to long term. Therefore, while the VSI Alliance moves to seed the emerging SOC industry, companies are developing intra-company reuse solutions not only to ratify what the VSI Alliance has proposed, but also to establish the infrastructure needed to change their way of doing design. What is being discovered is that even within the proprietary IP-friendly confines of a single company, reuse does not fit neatly into a tool, process, or technology. To experience productivity benefits from reuse requires having a system that addresses IP integration, creation, access, protection, value recognition, motivation, and support. *Until you have a viable IP system development plan, you do not know what to author, what to buy, what to redesign, what standards to use, or what barriers must be overcome (technical and non-technical).*

Reusing IP has long been touted as the fastest way to increasing productivity. Terms like "design factory" and "chip assembly" conjure up visions of Henry Ford-like assembly lines with engineers putting systems together out of parts previously designed in another group, in another country, in another company. Yet while this has been pursued over the past two decades at the highest levels of management in the electronic design and software development industries, we have only seen some small victories (for example, cell libraries, software object libraries) and a lot of unfulfilled promise. Why do we keep trying? Where do we fail?

Reuse does work, and when it works, it has spectacular results. At its most basic level, if an engineer or engineering team does something once and is then asked to do it or something similar again, a productivity increase is typically observed in the second pass. In this case, what is being reused is the knowledge in team members' heads as well as their experience with the processes, tools, and technology they used. However, if another engineer or engineering team is asked to execute the second pass, little productivity increase is observed. Why does this happen?

- Is this just a "not invented here" engineering viewpoint?
- A lack of adequate documentation and standards?
- Limited access to what has been done?
- The perception in design that learning and adapting what has been done takes longer than starting from a specification?

---

6. National Technology Roadmap for Semiconductors, August 1994; and National Technology Roadmap for Semiconductors, 1997, available at www.sematech.org/public/roadmap/index.htm.

- An unwillingness to accept a heavily constrained environment?
- An inability to create an acceptably constrained environment?
- A failure to see the difference between having someone use what has been done and having someone change what has been done to use it?

It is probably some or all of the above. But the impetus to overcome these barriers must come from inside both engineering and management.

The issue of IP reuse can be looked at in several ways. For it is in reuse that all of the technical, organizational, and cultural barriers come together.

Models of Reuse

This section defines four reuse models: personal, source, core, and VC. It also outlines the capabilities that are necessary to transition to the next stage of reuse. Figure 1.4 shows how the reuse models map to the TDD, BBD, PBD design methodology evolution. In our description of the reuse models, we use the term "portfolio" to represent the human talent and technological knowledge that pre-exists before attempting a new design project.

In the earlier phases, reuse is largely opportunistic and initiated at design implementation. As reuse matures and becomes part of the culture, it is planned and considered in the earliest phases of design and product planning, ultimately arriving at an infrastructure that supports full separation of authoring and integration.

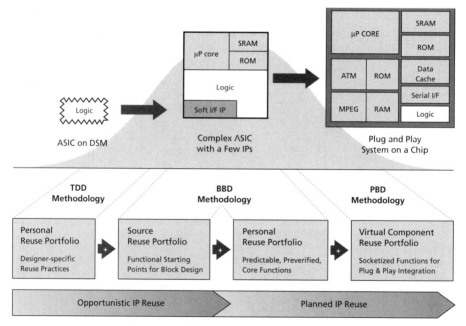

Figure 1.4. Evolution of VC Reuse

Personal Reuse Portfolio

In the traditional TDD ASIC methodologies, reuse is human knowledge-based, and is exercised through reapplying personal or team design experience to produce derivative projects.

Table 1.6. Benefits and Challenges of Personal Reuse

Benefits	Challenges
No specific preparation or standards required.	Applies only to single-threaded operations; does not scale.
Works well for single design team operations doing derivative products.	Dependent on retaining key personnel (exacerbated by best designers wanting to work on next system, not derivatives).
Tends to see reuse of design methodology and tools within the team model.	Changes to technology or architecture significantly undermine productivity benefits.

The transition from personal reuse to the next stage focuses mainly on infrastructure and laying the groundwork for future enhancements. Realizing the full potential for reuse requires solving certain technical and business issues. However, in many corporate environments, the biggest hurdle in the change process is overcoming the engineering tendency to invent at every opportunity.

Table 1.7 summarizes the functions and technologies that need to be in place to transition to source reuse. This first step gets people looking at existing designs from a reuse perspective and gives engineers an opportunity to identify the barriers to reuse. It also sends strong messages that reuse will be pursued both in investment funding and engineering time.

Table 1.7. Transitioning from Personal to Source Reuse

Function	Technology	Activity
Cataloging legacy and third-party VCs	BLOB (Binary Large Object) database; database schematic includes full model and documentation fields	Collect, filter legacy VCs; create access to third-party VCs; VC Web data
Assessment	RTL style guides, VSI Alliance format standards	Establish VC retention criteria; analyze legacy and third-party; edit database
Access: bi-directional to include feedback	Intracompany Web system; custom security procedures	Set up access system; establish process for update; establish feedback path to/from users

Source Reuse Portfolio

At the entry point into BBD, the opportunity to reuse designs created elsewhere begins to open up. The function, constraints, and instance-based context for a block are known as a result of a top-down system design process. Source reuse can speed up the start of a design by providing a pre-existing RTL or netlist-level design that can then be modified to meet the system constraints. The productivity benefits of this approach, however, are debatable depending on how well it matches the system constraints, its complexity, whether it is accompanied by an appropriate testbench, whether the original designer is available to answer questions, and the openness of the adopting designer to using an existing design. In addition to all these, the most significant barrier in many large companies is providing designers and design teams information on what is available in an accessible, concise form.

Table 1.8. Benefits and Challenges of Source Reuse

Benefits	Challenges
Can bootstrap a design start, depending on match to the system requirements.	Productivity is eroded if re-verification is required when the block is changed (and it almost always is).
Cataloging infrastructure eases access to internal IP and provides a foundation for future reuse investments.	Testbench development often limits productivity for many blocks.
Can be used to establish basic documentation guidelines for RTL and accompanying testbenches.	The time to evaluate and understand what is available often takes longer than writing RTL from specification.
Is an excellent stage to explore incentives for reuse and models for support of IP. Applies both on the creation and integration sides.	Predictability of performance, area, and power is poor, adding risk late in the project during physical implementation and integration.
Productivity factor* of .75x => 1.5x (largely a function of RTL documentation, testbench, access to support resources, and block complexity).	Most existing IP is not well documented for reuse, often lacking an adequate testbench.

*For purposes of comparison, the productivity factor is defined as the ratio of the time required to reuse an existing block (including modification and re-verification) to the time required to do an original design, given a set of block and system design specifications.

The transition from source to core reuse typically occurs in parallel with a maturing top-down, hierarchical, BBD methodology. The primary objective in this stage is to upgrade the VC portfolio to a level of documentation, support, and implementation that guarantees an increase in productivity. At the same time, the VC portfolio changes from general purpose, unverified VC source files to application-proven cores with a physical and project history. The system integrator can expect to find the following: more hard cores that are already implemented on the target technology; high-value anchor VCs that tend to dictate integration elements such as buses; VC cores that have been validated and tested before entry into the database; established and mature third-party IP relationships; and a set of models that are designed to be directly usable in the chip integration methodology and tools system. Table 1.9 summarizes the functions that need to be in place to transition to core reuse.

Table 1.9. Transitioning from Source to Core Reuse

Function	Technology	Activity
VC qualification	RTL linting tools; model cross-verifiers; coverage analyzers	VC qualification and characterization
VC block modeling and/or retargeting	Block design methods (tools and flows); model development tools (timing, power, floor plans, and behavioral); testbench development tools; technology porting systems	Legacy VC refurbishing to internal formats/standards; third-party VC customization to internal formats/standards
VC block design	Block design methods (tools and flows); model development tools; testbench development tools	Block authoring from specification and constraints

Among the technologies that play a key role in this transition are those that port the legacy VCs to a new technology. These include methods for soft, firm, and hard VCs, including extraction, resizing, and retargeting at the GDSII level. The retargeting ideally supports a performance/power/area optimization as an integral element in achieving the specific block objectives. The model development tools should be a natural outgrowth of block authoring design methods. These will be significantly enabled as the promised technology advances in timing characterization, RTL/block floor planning, and power characterization for real world DSM designs, with their non-trivial clocking schemes and state-dependent delay functions. Also required at this point are tools that support a VC qualification process as a screen for entry in the VC reuse database. In addition to conventional RTL linting tools, which are tuned to the intracompany VC sharing rules, coverage analyzers and model cross verifiers will establish a consistent level of VC quality. Block authors will also begin to demand model formal equivalence checking tools and methods to assure model versus implementation coherence. Finally, the development of a top-down, system-in, design verification strategy requires viable technologies for driving the results and tests from algorithmic, behavioral, and hardware/software co-verification testbenches through partitioning and onto the inputs and outputs of the reusable blocks.

Core Reuse Portfolio

As a design organization matures toward a hierarchical BBD methodology, a reusable VC portfolio is refined and improved in the following ways:

- More information is available on the block realization in silicon (area, timing, footprint, power).
- More blocks appear in firm or hard form.
- Reusable, placed netlist is in a relevant technology library (firm).
- Qualification constraints exist for entering into the reuse VC database and pruning existing low-value entries.
- There is documented data on the context in which the block has been used and/or reused.
- Third-party VC blocks are integrated, and specific details on the engagement process with the vendor exist.
- More participation and mechanisms supporting the block occur from the author for internal VCs as a result of refinements to the incentive program.

At this point, the combination of a mature BBD methodology, increased reuse, and management/market pressures tends to break down some of the non-technical reuse barriers. Other benefits include: the availability of larger, more complex blocks, often as GDSII hard cores; the use of high-level models (above RTL) as a consequence of more top-down design methods; and the formation of design teams consisting of system designers, chip integrators, and block authors. In addition, testbenches for the blocks are derived more from system-level tests than from independent development.

The final transition creates a VC portfolio that supports a true plug and play environment. This means separating authoring and integration in a practical fashion by designing the key integration platform architectures and incorporating their implicit virtual system constraints into the IP authoring process. New tools emerge that address developing an integration platform into which selected pre-characterized and preverified VCs can be plugged without modification.

Table 1.10. Benefits and Challenges of Core Reuse

Benefits	Challenges
Basic organizational barriers to reuse are bridged.	The VC cores must be modified to adapt to the bus, clock, test, and power environment of the product design, which takes time and introduces errors.
Documentation quality is sufficient for realizing an observable productivity benefit.	Verification of blocks in the system context after modification is still a problem.
Methods for "resurrecting and refurbishing" legacy IP are emerging.	Architectures for dealing with mixed-signal design are ad hoc in terms of block reuse across products.
Top-down BBD methodology is in place. Reuse factor of 1.5x to 3x.	The productivity benefits are not growing at the same rate as the design size and DSM interconnect problems.

With the move toward preverified components, verification shifts to an interface-based focus in which the base functionality is assumed to be verified at the system level. Because more VCs are likely to be hard, VC migration tools will mature, providing a port across technology transitions. In addition to all performance sensitive VCs being pre-staged in silicon, soft or firm VCs will be provided in an emulation form for high-performance hardware/software verification.

Table 1.11. Transitioning from Core to VC Reuse

Function	Technology	Activity
VC qualification	VC characterization methodologies; model cross verifiers; coverage analyzers	VC prequalification and characterization; VSI-compliant service providers; VC pre-staging in silicon
VC authoring and socketization	VC authoring guides (tools and flows); model development tools (firm/hard); VC migration tools; interface-based testbench development systems; documentation	Legacy VC refurbishing to VC standards; third-party VC customization to internal VC model development; VC porting; VC block authoring for reuse from specification and virtual system architectural specification
VC delivery vehicles	Virtual system design tools (bus analyzers, signal integrity analyzers, substrate models); interface-based verification testbench development systems	Co-pilot and shadow project; process optimization; delivery vehicle development

VC Reuse Portfolio

The transition of IP into a VC status is where the greatest productivity benefits are realized and where the separation of authoring and integration is most clearly observed. These VCs are pre-characterized, preverified, pre-modeled blocks that have been designed to target a specific virtual system environment. This virtual system design, which consists of a range of operational constraints bounded by performance, power, bus, reliability, manufacturability, verification characteristics, cost, and I/O, is applied to a specific market/application domain. This reuse domain includes the functional blocks, each block's format and flexibility, the integration architecture into which the blocks will plug, the models required for VC evaluation and verification, and all of the constraints to which the blocks must conform.

Within the application domain, the IP is differentiated largely by the elegance of the design, the completeness of the models/documentation, and the options presented within the domain context. The bus speed and protocol can be given, but the area/power/speed ratio and supported bit width are often variable. The test architecture might dictate that blocks be either full scan or full BIST with a chip level JTAG, but IP will be differentiated on the coverage to vector ratio and even the failure analysis hooks provided. The VCs are pre-staged and qualified before being added to the environment. Because the blocks are known entities, designed to fit together without change, the productivity

Table 1.12. Benefits and Challenges of VC Reuse

Benefits	Challenges
VCs designed to be used as is, very often firm or hard, tuned to domain power/performance/cost, architectural, and manufacturability targets.	TTM forces trade-offs in flexibility and optimality.
VC portfolio(s) focused on specific high-leverage market/application domain platforms.	Software VCs begin to assume an expanding role. More high-performance analysis and modeling needed to support hardware/software partitioning decisions.
Preverified VCs modeled for more manageable interface-based design verification (vs. exponentially exploding combined function and interface verification).	Intracompany VC and SOC methodologies dominate. Fundamental business models for VC control and protection are a barrier to intercompany VC sharing. Alliances and partnerships provide vehicles for establishing critical mass of VC.
Virtual architectures defined and implemented for high-leverage market domains. Targeted mixed-signal chip integration architectures emerging.	Mixed-signal solutions dependent on process as well as design architectures. Porting analog cores still manual, but now targeted to mixed-signal integration architectures.

from this type of environment can increase more than 10x. The penalties in area or performance to get such TTM benefits are less than one would anticipate because of the optimization of the domain-focused VC portfolio to the technology and manufacturing constraints.

Developing an Integration-Centric Approach

In addition to adopting a VC reuse portfolio, a different perspective is needed to realize the necessary productivity increase required to address TTM and design realities. To achieve new solutions, we need to look at the issues from an integration-centric perspective rather than an IP-centric one, as summarized in Table 1.13.

Some of the steps that need to be taken to implement an integration-centric approach for reuse are as follows:

1. *Narrow the design focus to a target application family domain.* The scope of the application domain is a business decision, tempered by the technical demands for leverage. The business issues center around product market analysis, derivative product cycles, possible convergence of product lines,

Table 1.13. IP-Centric vs. an Integration-Centric Approach

IP-Centric Issues	Integration-Centric Issues
How can I create IP that can be modified for reuse in all applications?	How can I create IP that can be reused without change?
What IP do I have? How can I make it known?	What IP do I need? What do I have to own?
How can I create an IP portfolio large enough to address any market?	What are the key markets I want to serve? What IP do they need? How will it be used?
How can I make my IP portfolio like my ASIC cell library, with common flows to match?	How can I create a system in which the IP and the methodologies are optimized for a target market?
How can I design IP for flexibility and re-verification after I have adapted the IP to the application?	How can I pre-stage the IP for the target application so that it plugs into my integration platform?
How can I adapt the IP to the product requirements of each project?	How can I adapt the IP to a product domain so that the architecture and IP do not need to be changed?
The enemy of time to market is inflexibility!	**The enemy of time to market is change!**

and the differentiating design/product elements that distinguish the products. Where TTM and convergence of applications or software algorithms are critical differentiators, moving to a VC reuse portfolio and PBD methodology is essential.

2. ***Identify the VC blocks that are required for each domain.*** Separate the VCs as follows:
   - differentiating and needing control
   - acquired or available in the current market
   - internal legacy IP

3. ***Develop a virtual system design for the target platform*** that identifies the VC blocks, the integration architecture, the block constraint ranges, the models required, and the design and chip integration methods to author, integrate, and verify the product design. Extend your guideline-oriented documentation to comprehensive VC authoring and integration guides that include processes, design rules, and architecture.

4. ***Socketize/productize your application domain VCs*** to conform to the virtual system constraints. This includes determining multiple implementations (low power, high performance), soft versus firm versus hard, and creating and verifying high-level models. Depending on the function, it also includes preverifying the core function and isolating the interface areas for both verifying and customizing. To achieve the optimal value, all performance critical VCs should be pre-staged and fully characterized in the target silicon technology, much the same way you would do with a cell library. For verification purposes, pre-staging the VC for a field-programmable gate array (FPGA)-type prototyping/ emulation environment (for example, Aptix or Quickturn) is also recommended for any VC that is involved in a subjective or high-performance verification environment.

5. ***Demonstrate and document the new application environment*** on a pilot project to establish that the virtual architecture, authoring and integration methods, and software environments are ready. This also identifies the refinements necessary for proliferating the new design technology across the organization.

6. ***Optimize the authoring and integration design processes and guides*** based on the pilot experience. This is an ongoing process as technology and market characteristics evolve.

7. ***Proliferate the platform across the organization.*** Using both the momentum and documentation from the pilot, deploy measurement methods that will allow the productivity benefits to be tracked and best practices identified.

SOC and Productivity

Many elements must come together to achieve SOC. The device-level technologies necessary to support the silicon processing evolution to sub-0.2 micron designs through manufacturing and test are still being defined. The design flows and tools for authoring and chip integration are immature and, in some cases, still to be delivered. However, two key design technologies are emerging to address the productivity side of SOC. These technologies, integration platform and interface-based design, represent an amalgam of design principles, tools, architectures, methodologies, and management. But before we explore these key technologies, we will look at the steps and tasks involved in SOC design.

2

Overview of the SOC Design Process

Creating a systematic environment is critical in realizing the potential of SOC design. Figure 2.1 depicts the basic elements of such an environment. This chapter describes each of these areas briefly and in the context of platform-based design; subsequent chapters will discuss the steps and elements involved in platform-based design in more detail. Figure 2.1 will also be used in each chapter

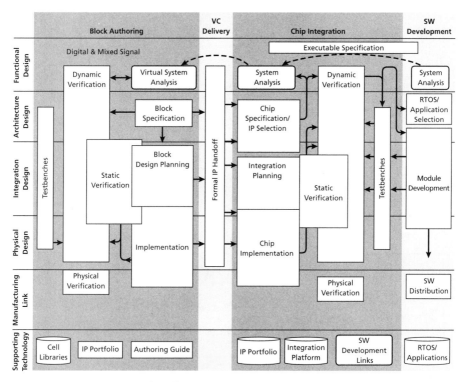

Figure 2.1. Basic Elements of Platform-Based Design

to highlight which areas of platform-based design we are addressing. These diagrams are not intended to be viewed as design flow diagrams, but rather as high-level maps of the SOC design process so that each element in the process can be seen in relation to the others.

Block Authoring

Figure 2.2 provides a more detailed view of the block authoring process. The role of each element is described below.

Rapid Prototyping

Rapid prototyping is a verification methodology that utilizes a combination of real-chip versions of intellectual property (IP) blocks, emulation of IP blocks in field-programmable gate arrays (FPGA) (typically from a register-transfer level (RTL) source through synthesis), actual interfaces from the RTL, and memory to provide a very high-speed emulation engine that permits

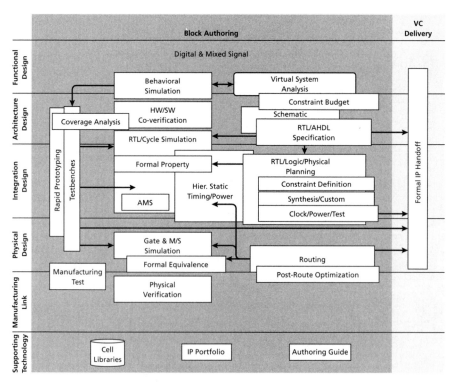

Figure 2.2. Detailed View of Block Authoring

hardware/software verification using real-world verification sources, such as video streams. This technology is effective in many environments, especially where hardware versions of processing elements are available. The amount of design put into FPGAs is practically bound, and the medium for verification is subjective in nature, for example, picture quality. Using rapid prototyping in the authoring context is generally limited to two applications:

- Insuring that the block can be synthesized to an FPGA library (when an RTL source is supplied)
- Verifying that blocks can handle real-world video, audio, or high bandwidth data streams

Testbenches

Testbenches are tests, for example, stimulus–response, random seed, and real-world stream, run at all levels of design from system performance to gate or transistor level, that are used to verify the virtual component (VC). Segmenting the tests by simulator format and application is expected, and subsetting tests for power analysis, dynamic verification of timing escapes, or manufacturing is appropriate. Modern testbenches are very much like RTL or software design and should be documented accordingly for ease of understanding. The quality of the testbench is a fundamental barometer for the quality of the VC. Information detailing the testbench coverage over the VC design space is becoming a differentiator for integrators when selecting among functionally equivalent VCs.

Coverage Analysis

Coverage analysis is a design task that analyzes activity levels in the design under simulation to determine to what degree the testbench verifies the design functionally. A variety of tools that provide a numerical grade and point to the areas of the design that are poorly covered support this task. In some cases, they suggest ways to improve the design. Tools also used during coverage analysis are RTL "linting" tools, which determine whether the RTL satisfies style and documentation criteria. These linting tools are sometimes used for determining whether a block should be included in the corporate IP library.

Hardware/Software Co-verification

Hardware/software co-verification verifies whether the software is operating correctly in conjunction with the hardware. The primary intention of this simulation is to focus on the interfaces between the two, so bus-functional models can be used for most of the hardware, while the software would run on a model of the target CPU. In this manner, the software can be considered part of the testbench for the hardware. Because of the many events necessary to simulate software operations, this level of simulation needs to be fast, thus requiring high-level models or prototypes. Often only the specific hardware logic under test will be modeled at a lower, more detailed level. This allows the specific data and events associated with the hardware/software interface to be examined, while speeding through all of the initialization events.

Typically, the focus of the hardware/software co-verification is to verify that the links between the software and hardware register sets and handshaking are correct, although this co-verification can go as far as verifying the full functionality, provided that there is sufficient simulation speed. In the block authoring role, this step is used to verify the combined hardware/software dual that comprises complete programmable VCs. The software drivers can also be matched to several different RTOSs and embedded processors.

Behavioral Simulation

Behavior simulation is based upon high-level models with abstracted data representation that are sufficiently accurate for analyzing the design architecture and its behavior over a range of test conditions. Behavioral models can range from bus-functional models that only simulate the block's interfaces (or buses) accurately to models that accurately simulate the internal functions of the block as well as the interfaces. The full functional models can also be timing-accurate and have the correct data changes on the pins at the correct clock cycle and phase, which is called a cycle-accurate model. Behavioral simulation is slower than performance simulation, but fast enough to run many events in a short period of time, allowing for entire systems to be verified. The test stimulus created with behavioral simulation and its results can be used to create the design testbench for verifying portions of the design at lower abstraction levels. The SOC environment also demands that these models include power consumption parameters, often as a function of software application type. The behavior of these models must be consistent with the VC implementation and apply to analog as well as digital blocks.

RTL/Cycle Simulation

This simulation environment is based primarily upon RTL models of functions, allowing for cycle-based acceleration where applicable, as well as gate/event-level detail where necessary. Models should have cycle accurate, or better, timing. The intended verification goal is to determine that functions have been correctly implemented with respect to functionality and timing. Testbenches from higher level design abstractions can be used in conjunction with more detailed testbenches within this simulation environment. Since this is a relatively slow simulation environment, it is typically only used to verify and debug critical points in the circuit functionality that require the function and timing detail of this level. As such, this environment can be used as the debug and analysis environment for other verification technologies, such as rapid prototyping and formal methods.

Formal Property Checking

Formal property checking tools can be an efficient means to verify that the bus interface logic meets the bus protocols and does not introduce erroneous conditions causing lock-up or other failed conditions.

This task involves embedding assertions or properties that the design must satisfy in the hardware description language (HDL). The formal property checker uses formal methods to determine whether the detailed RTL design satisfies these properties under all conditions. Although useful in many applications, such as cache coherency and state machines, this verification technique is limited to the more sophisticated VCs in the authoring space.

Analog/Mixed Signal (AMS)

This task recognizes that many analog blocks are in fact digital/analog mixed-signal hybrids. Using an analog hardware description language (AHDL) that is amenable to mixed simulation at the RTL level is critical for properly analyzing large mixed-signal blocks and capturing design intent for reusing AHDL and the schematic as a starting point for technology migration.

Hierarchical Static Timing/Power Analysis

Static timing analysis is emerging as a sign-off quality methodology for today's semiconductor technologies. This method is very amenable to hierarchical and authoring-based block designs, provided that methods are put in place for handling issues such as state dependent delays, off-block loading, clocking schemes, and interfaces to asynchronous behaviors. A link to the event simulator is needed for design elements that require dynamic verification. Power models are emerging much in the same fashion as static timing. However, power calculations require that a model of the node transition activity be developed, typically from a subset of the simulation testbench, and, hopefully, one that is based on analysis of system-level behavior. Once the frequency is established, calculating the power consumption of a block based on the GDSII level of implementation and power supply voltage is feasible. In addition to verifying that the block has satisfied the pre-established goals for implementation, this function also outputs the Virtual Socket Interface (VSI)-compliant models for the block across the continuum of behavioral ranges, which include clock frequencies and voltage ranges.

Gate and Mixed-Signal Simulation

Although gate and mixed-signal simulation are shown as a single task in the diagram in Figure 2.2, they are actually two separate disciplines. Gate-level digital simulation is still used in some situations for timing verification and to ensure that the RTL testbench runs after the design has been manipulated through synthesis, buffering, and clock and I/O generation. For non-synchronous digital design, this method is used to ensure that the timing and power constraints have been met. A model generation technique is needed to provide a timing model for the block user. In the mixed-signal domain, a high-performance device-level simulator provides both the functional verification and the power/timing verification. Again, a model generator for interfacing up the hierarchy is required.

Formal Equivalence Checking

Equivalence checking tools verify on a mathematical basis, without testbenches, that the gate-level netlist and the RTL are functionally equivalent. Differences, when detected, need to be linked back to the simulation environment for analysis and debugging.

Physical Verification

Physical verification includes extracting from the final layout a transistor model and then determining whether the final design matches the gate-level netlist and meets all the electrical and physical design rules. The introduction of a block-level hierarchy to ensure that chip-level physical verification proceeds swiftly requires that the block model be abstracted for hierarchical checking.

Manufacturing Test

Generating an appropriate set of test vectors and test access mechanisms to test the correct manufacturing of the part (not the correct implementation of the specification) is a fundamental element for all blocks. For soft blocks, where test insertion is assumed, a demonstration of test coverage is appropriate to ensure observability and controllability. This function creates a test list for the block, provides a coverage figure along with any untestable conditions or areas, and documents any test setup requirements. This is required for all blocks, digital and analog. Standards such as IEEE 1149.1, 1149.4, P1450, and P1500 are applicable.

Virtual System Analysis

When designing a block for reuse, first the design function needs to be determined, followed by what are the design targets, constraints, external interfaces, and target environments. Normally, such constraints and contexts come from the product system design. However, when the goal is a block for reuse, the block author must execute a virtual system design of the target reuse market to know what standards and constraints must be met. And since the block's reusability needs to span a range of time and designs, often the constraints are expressed in ranges of clock frequency or power consumption or even interface standards. The process for this is most similar to normal product system design with derivatives planning, but requires a broader understanding of the target application markets.

Constraint Budget

The constraint budget, which is a result of the virtual system design, describes the design requirements in terms of area, power, and timing—both discrete and ranged—over which the block must perform. Also included are things like test coverage and time on tester, industry hardware/software benchmarks, bus interfaces, test access protocols, power access protocols, design rules for noise isolation, and the list of all models needed.

Schematic

The schematic capture of the design is primarily used for analog blocks. It is also used for capturing high-level design through block diagrams, which are then translated into structural HDLs.

RTL/AHDL Specification

RTL is the primary hardware design language implementation model for digital designs. The RTL specification describes transformation functions performed between clocked state capture structures, such as registers. This allows the functions to be synthesized into Boolean expressions, which can then be optimized and mapped onto a technology-specific cell library. A variety of rules have evolved for writing RTL specifications to ensure synthesizability as well as readability. These rules should be implemented; tools are available for checking conformance to rules (see "Coverage Analysis" on page 31). Two languages used today are VHDL and Verilog.

AHDL is the analog equivalent of RTL. The state of analog synthesis is such that the HDL representation correlates to a design description that is suitable for functional simulation with other blocks (digital and analog), but from which the actual detailed design must be implemented manually (no synthesis). AHDL does meet the documentation requirements for intent, and coupled with the schematic of the circuit, serves as a good place for migrating the design to the next generation technology.

RTL/Logic/Physical Planning

This activity is the control point for the physical planning and implementation of the block. Depending on the block type (soft, firm, or hard) and complexity, floor planning can play an important role in the authoring process. Planning considers and assigns, as needed, I/O locations, subblock placement, and logic to regions of the block. It eventually directs placement of all cells in the target or reference technology. It invokes a variety of specific layout functions, including constraint refinement, synthesis or custom design, placement, clock tree, power, and test logic generation. Fundamentally, the power of the planner lies in its ability to accurately predict the downstream implementation results and then to manage the interconnect through constraints to synthesis, placement, and routing. To do this, the planner must create a predictive model starting at RTL that includes the critical implementation objectives of area, power, and performance. The minimization of guardband in this area enables the author to differentiate the block from other implementations. Firm IP block types can carry the full placement or just the floor plan forward to give the end user of the block the most predictive model. The model for performance and the power produced is captured and used in the IP selection part of the assembly process.

Constraint Definition

The original constraints for the block represent the system-level requirements. During design implementation, these constraints are refined and translated to detailed directives for synthesis, timing, placement, and routing.

Synthesis and Custom Implementations

For digital design, two styles of detailed implementation are generally recognized: synthesis from the RTL, and custom netlist design at either the cell or transistor level. There are many combinations of these two techniques that have been deployed successfully. The synthesis style relies on a cell library and directives to the synthesis tool to find a gate-level netlist that satisfies the design requirements. Placement is iterated with synthesis and power-level adjustments until an acceptable result is achieved. Test logic and clock trees are generated and can be further adjusted during routing. This style is very efficient and effective for up to moderately aggressive designs. Some weaknesses show when the design is large and must be partitioned, or when the design has an intrinsic structure, such as a datapath that the synthesis tool is unable to recognize. The increasing dominance of wires in the overall performance and power profile of a VC is dictating that placement and synthesis need to be merged into a single optimizing function rather than an iterative process.

Very aggressive design techniques, such as domino logic or special low-power muxing structures, typically require a more custom approach. The custom approach typically involves augmenting the cell library with some special functions, then using advanced design methods where the logic and the physical design (including clock and test) are done as a single, unified design activity. For very high-value IP, where area, performance, and power are all optimized, such as processors, custom implementations are the norm. All analog blocks are custom designs.

Clock/Power/Test

The insertion of clock tree, power adjustments, and test logic can be done at any time in the process, although usually they are added once the placement of the logic has settled. Global routing, which is invoked at several levels depending on the degree of hierarchy in the block, takes the clock tree into consideration. Automated test logic insertion is the most conventional approach and utilizes scan-based test techniques. Built-in-self-tests (BIST), which are design elements built into the design, generate test vectors based on a seed and then analyze a signature to determine the result. It is vital that whatever approach is used for the block, the test access protocol be clearly defined.

Routing

The routing of a block is always done for hard implementations and responds to the constraints from the planning phase in terms of I/O, porosity by layer, clock skew, and power buffering. The router is becoming the key tool for deep submicron (DSM) designs for dealing with cross talk, electromigration, and a host of other signal integrity issues that arise as geometries shrink and new materials are introduced. Routers that are net to layer selective, able to provide variable line width and tapering, able to understand the timing model for the system, and react to delay issues dynamically are emerging as the tools of choice.

Post-Routing Optimization

Some of the routing optimization options available have improved the area/power/performance trade-offs by significant amounts (>10 percent). These transistor-level compaction techniques adjust power levels to tune the circuit for performance. This reduces area and either power or performance. Some authors of hard IP will choose to provide a block with multiple profiles, one for low power, another for high performance. For these authors, post-routing techniques are very useful. Similarly, when implementing a soft block, chip integrators can take advantage of post-routing optimizations to get a difficult block into their constraint space.

Cell Libraries

Cell libraries are the low-level functional building blocks used to build the functional blocks. They are typically technology-specific, and contain many different views (such as logic models, physical layout, delay tables) to support the steps in the design flow.

Authoring Guide

The authoring guide is the design guide for IP block authors that specifies the necessary outputs of the block authoring process, as well as the design methodology assumptions and requirements of the chip integration process so that the blocks will be easily integrated. These requirements would include documentation, design, tool environment, and architecture requirements.

IP Portfolio

The IP portfolio is the collection of VC blocks that have been authored to meet the authoring guide requirements and which meet the design goals of the set of VC designs that an integration platform is intended to serve. VC blocks within a portfolio are tailored to work with a specific integration platform to reduce the integration effort, although some VC blocks might be general enough to be a part of multiple portfolios and work with multiple integration platforms.

VC Delivery

Figure 2.3 provides a more detailed view of the VC delivery process. The role of each element is described below.

Formal VC Handoff

This is the VSI-compliant, self-consistent set of models and data files, which represent the authored block, that is passed to the chip integrator. The design tools and processes used to create the VC models, and the design tools and processes used to consume the VC models and build a chip, must have the same semantic understanding of the information in the models. Tools that claim to read or write a particular data format often only follow the syntax of that format, which might result in a different internal interpretation of the data, leaving the semantic differences to be discovered by the tool user. Without methodologies that are linked and proven semantically, surprises between author and integrator can arise to the detriment of the end product.

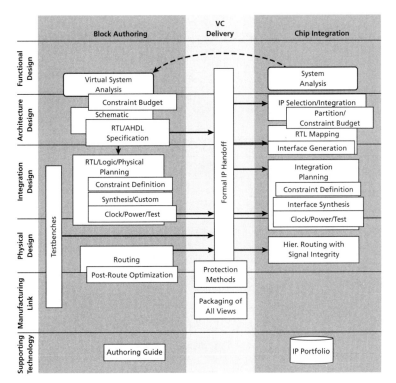

Figure 2.3. Detailed View of VC Delivery

Protection Methods

There are a variety of encryption methods proposed by OMI and other standards bodies. In addition to encryption, there are methods proposed for watermarking the design to permit tracing VCs through product integration. As these methods and the legal system for exchanging VCs matures, tools for providing first-order protection of VCs will be commonplace.

Packaging of All Views

Maintaining a self-consistent set of views for VCs, with appropriate versioning and change management, is anticipated to be an essential element in a VC delivery package. This will enable both the author and the integrator to know what has been used and what has been changed (soft VCs are highly subject to change).

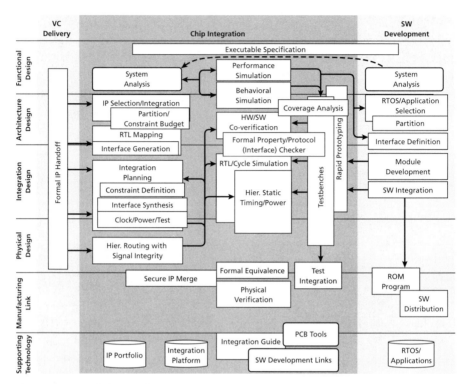

Figure 2.4. Detailed View of Chip Integration

Chip Integration

Figure 2.4 provides a more detailed view of the chip integration process. The role of each element is described below.

Executable Specification

An executable specification is the chip or product requirements captured in terms of explicit functionality and performance criteria. These can be translated into constraints for the rest of the design process. Traditionally, the specification is a paper document, however, by capturing the specification as a formal set of design objectives, using simulatable, high-level models with abstract data types and key metrics for the design performance, the specification can be used in an interactive manner for evaluating the appropriateness of the specification itself and testing against downstream implementation choices. These models are typically written in C, C++, or SDL.

System Analysis

System analysis develops and verifies the algorithmic elements in the design specification. These algorithms are the base for fundamental partitioning between hardware and software, meeting the first-order constraints of the specification, such as application standards and selecting the target technology for implementation.

VC Selection/Integration

VC selection includes the evaluation of both the blocks available and the platform elements. In platform-based design, many of the essential elements of the platform are in place and the selection process consists largely of making refinements that might be needed to meet the system requirements. For new blocks that have not yet been developed, the interface requirements and functional behavior are defined.

Partition/Constraint Budget

The hardware elements are partitioned, and detailed performance, power, and interface constraints are defined. At this stage, the implementation technology is assessed against the partitions and the integration design, and an initial risk guardband, which identifies areas where the implementation will need particular attention, is developed. The architectural elements for power, bus, clock, test, and I/O are all put into place, and blocks that must be modified are identified. This process is likely to be iterative with the VC selection, especially where significant new design or platform changes are contemplated.

RTL Mapping

The design is mapped into a hierarchical RTL structure, which instantiates the VC blocks and the selected platform elements (bus, clock, power), and kicks off the block modifications and new block design activities.

Interface Generation

The RTL design modifications are entered to blocks that require manual interface modification. For blocks designed with parameterized interfaces, the parameters that will drive interface logic synthesis are established. Interfaces include all architectural as well as I/O infrastructures.

Integration Planning

The integration planner is the vehicle for physically planning in detail the location of the blocks, the high-level routing of the buses and assembly wiring, considerations regarding clock trees, test logic, power controls, and analog block location/noise analysis, and the block constraints based on the overall chip plan. The blocks with very little guardband tolerance are either adjusted or queued for further analysis.

Constraint Definition

The constraints are detailed based on the floor plan and used to drive the final layout/route of the integration architecture. Critical chip-level performance paths and power consumption are analyzed, and constraints are adjusted to reflect the realities of interconnect wiring delays and power levels extracted from the floor plan.

Interface Synthesis

Blocks that are set up for auto synthesis of interfaces are "wrapped" to the integration architecture.

Clock/Power/Test

This task generates the clock trees for both digital and analog, lays down the power buses and domains, taking into account noise from all sources (digital to analog isolation, ground bounce, simultaneous switching, and so on), and inserts the test logic.

Hierarchical Routing with Signal Integrity

This step is the final routing of the block to block interconnect and the soft VC blocks, which can be floor planned into regions. Hierarchical routing requires a hybrid of high-level assembly and area-routing techniques that share a common understanding of constraints and signal integrity. The final detailed delays and power factors are extracted and fed into the analysis tools. The correlation of assumptions and assertions made all the way up in the VC selection phase to actual silicon will be progressively more difficult as more complex designs (faster, larger, mixed analog and digital) are implemented on single chips in DSM technologies. The signal integrity issues alone require a truly constraint-driven routing system that adapts the wiring and the active elements to the requirements.

Performance Simulation

Performance simulation is based on high-level models that have limited representation of functionality detail, but are intended to provide high-level performance estimates for evaluating different implementation architectures. This simulation environment can be used to make algorithm selection, architectural choices, such as hardware versus software partitioning trade-offs, and VC selection. It also provides estimates on the feasibility of the design goals. The simulation is usually of models that represent the critical path or key functional mode for the hardware or software, which can be code fragments that are running on high-level models of key CPU functions. Performance simulation is very fast, because it is limited to functional detail; therefore, it can be used to evaluate many architectural variations quickly. The performance simulation step can be part of systems analysis.

Behavioral Simulation

Behavior simulation is based upon high-level models with abstracted data representation that are sufficiently accurate for analyzing the design architecture and its behavior over a range of test conditions. Behavioral models can range from bus-functional models that only simulate the block's interfaces (or buses) accurately to models that accurately simulate the internal functions of the block as well as the interfaces. The full functional models can also be timing-accurate and have the correct data changes on the pins at the correct clock cycle and phase, which is called a cycle-accurate model. Behavioral simulation is slower than performance simulation, but fast enough to run many events in a short period of time, allowing for entire systems to be verified. The test stimulus created with behavioral simulation and its results can be used to create the design testbench for verifying portions of the design at lower abstraction levels.

Hardware/Software Co-verification

Hardware/software verifies whether the software is operating correctly in conjunction with the hardware. The primary intention of this simulation is to focus on the interfaces between the two, so bus-functional models can be used for most of the hardware, while the software would run on a model of the target CPU. In this manner, the software can be considered part of the testbench for the hardware. Because of the many events necessary to simulate software operations, this level of simulation needs to be fast, thus requiring high-level models. Often only the specific hardware logic under test will be modeled at a lower, more detailed level. This allows the specific data and events associated with the hardware/software interface to be examined, while speeding through all of the initialization events.

Rapid Prototyping

Rapid prototyping in the chip integration phase is critical for verifying new hardware and software design elements with existing VCs in the context of the integration platform architectural infrastructure. Although the operational characteristics are as defined for block authoring, during chip integration this task focuses on verifying the high performance of the design's overall function. Significant advantage is achieved where bonded-out core versions of the VCs are available and interfaced to the bus architecture. In some situations, the rapid prototype can provide a platform for application software development prior to actual silicon avalibility, which can significantly accelerate time to market.

Formal Property/Protocol (Interface) Checker

Formal property checking tools can be an efficient means to verify that the bus interface logic meets the bus protocols and does not introduce erroneous conditions causing lock-up or other failed conditions.

Again, this ought to tie in with an executable interface specification where protocols can be "policed" for illegal actions.

RTL/Cycle Simulation

This simulation environment is based primarily upon RTL models of functions, allowing for cycle-based acceleration where applicable, as well as gate/event-level detail where necessary. Models should have cycle-accurate, or better, timing. The intended verification goal is to determine that functions have been correctly implemented with respect to functionality and timing. Testbenches from higher level design abstractions can be used in conjunction with more detailed testbenches within this simulation environment. Since this is a relatively slow simulation environment, it is typically only used to verify and debug critical points in the circuit functionality that require the function and timing detail of this level.

Hierarchical, Static Timing, and Power Analysis

Static timing analysis provides a comprehensive verification of the design's timing behavior by accumulating delays on all valid logic paths in the design. This is used to confirm that all timing goals and constraints are met. This method can then be applied hierarchically, where path delays and timing constraints can be calculated for a block and then represented on the top-level pins of the block. Determining which valid logic paths to calculate is a design challenge that often requires dynamic simulation; hence, static timing analysis and simulation are complementary verification methods. Power analysis is another complementary verification step that can be calculated on a hierarchical basis. Power calculation is most significantly influenced by circuit switching, so an estimation of switching is necessary, either from dynamic simulation or from estimations based upon clock rates.

Coverage Analysis

Coverage analysis tools can be used to determine how effective or robust the testbenches are. They can determine the number of logic states tested by the testbench, whether all possible branches in the RTL have been exercised, or other ways in which the intended or potential functionality of the design has been tested.

Formal Equivalence Checking

Formal equivalence checking uses mathematical techniques to prove the equivalence of two representations of a design. Typically, this is used to prove the equivalence of a gate-level representation with an RTL representation, thus validating the underlying assumption that no functional change to the design has occurred.

Physical Verification

Physical verification includes extracting from the final layout a transistor model and then determining whether the final design matches the gate-level netlist and meets all the electrical and physical design rules. The introduction of a block-level hierarchy to ensure that chip-level physical verification proceeds swiftly requires that the block model be abstracted for hierarchical checking.

Test Integration

Generating a cost-effective set of manufacturing tests for an SOC device requires a chip-level test architecture that is able to knit together the heterogeneous test solutions associated with each block. This includes a mechanism for evaluating what the overall chip coverage is, the estimated time on the tester, the pins dedicated for test mode, Test Access Protocol (TAP) logic and conventions, collection methods for creating and verifying a final chip-level test in the target tester environment, performance-level tests, procedures for test screening, generation and comparator logic for self-test blocks, and unique access for embedded memory and analog circuits. In addition, there are a number of manufacturing diagnostic tests that are used to isolate and analyze yield enhancement and field failures.

Secure IP Merge

The techniques used for IP protection in the block authoring domain will require methods and software for decoding the protection devices into the actual layout data when integrating into the chip design. Whether these take the form of keys that yield GDSII creation or are placeholders that allow full analysis prior to integration at the silicon vendor, the methodology will need to support it.

Integration Platform

An integration platform is an architectural environment created to facilitate the design reuse required to design and manufacture SOC applications and is often tailored to specific applications in consumer markets. Chapter 3 discusses integration platforms.

Integration Guide

The integration guide specifies the design methodology, assumptions, and requirements of the chip integration process. It also covers the design styles, including specific methodology-supported techniques, the tool environment, and the overall architecture requirements of the chip design.

IP Portfolio

The IP portfolio is a collection of VCs, pre-staged and pre-characterized for a particular integration architecture. An IP portfolio offers the integrator a small set of choices targeted for the product application domain under design.

Software Development Links

The relationship between hardware and software IP can often be captured in hardware/software duals, where the device driver and the device are delivered as a preverified pair. By providing these links explicitly as part of the platform IP, the integrator has less risk of error, resulting in more rapid integration results.

PCB Tools

The printed circuit board (PCB) tools must link to the chip integration process in order to communicate the effects of the IC package, bonding leads/contacts, and PCB to the appropriate IC design tools. Likewise, the effects of the IC must be communicated to the PCB tools.

Software Development

Figure 2.5 provides a more detailed view of the software development process. The role of each element is described below.

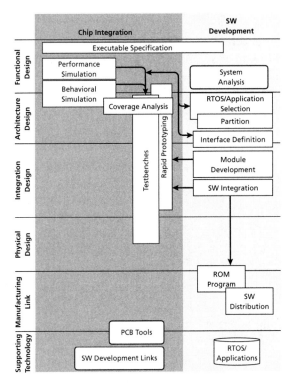

Figure 2.5. Detailed View of Software Development

Systems Analysis

Systems analysis is the process of determining the appropriate algorithm, architecture, design partitioning, and implementation resources necessary to create a design that meets or exceeds the design specification. This process can leverage design tools and other forms of analysis, but is often based heavily upon the experience and insight of the entire product team. The systems analysis of the software and hardware can occur concurrently.

RTOS/Application Selection

In this design step, the software foundations to be used to create the design, if any, are selected. The RTOS, the key application software, or other software components can significantly influence the structure of the rest of the software system. It needs to be selected early on, and might be part of the systems analysis/performance simulation evaluation process.

Partitioning

Partitioning determines the appropriate divisions between the functional elements of the design. These divisions are based on many factors, including performance requirements, ease of design and implementation, resource allocation, and cost. The partitions can be across hardware/software, hardware/hardware, and software/software. Systems analysis and performance simulation can facilitate this step.

Interface Definition

Once the functional elements, or modules/blocks, have been partitioned, the next step is to define the appropriate interfaces and interface dependencies between the elements. This step facilitates parallel or independent development and verification of the internal functions of the modules. This step can utilize performance or behavioral simulation to validate the interface architecture.

Module Development

Module development is creating the internal logic of the partitioned elements. The modules are created and verified independently against the design specification using many of the verification abstraction levels and design testbenches.

Software Integration

Software integration is the process of developing the top-level software modules that connect the lower-level software modules and the appropriate links to the hardware.

ROM/Software Distribution

Once the software component of the design has been completed and verified independently as well as in conjunction with the hardware, it can be released to production. This can take any variety of forms, from being written into a ROM or EPROM for security reasons, to being downloaded to RAM or flash memory from a network, the Internet, or from disk media.

RTOS and Application Software

The RTOS and application software provide the software foundation on which the software drivers and modules are built.

Design Environment

An integration platform serves as the basic building block for efficient SOC design. It is a pre-designed achitectural environment that facilitates reuse for the design and manufacturing of SOC applications in consumer-driven markets. The next chapter provides an overview of the different levels of platforms and how they are used depending on the target market.

3

Integration Platforms and SOC Design

The term platform in the context of electronic product design has been applied to a wide variety of situations. There are hardware platforms, software platforms, PC platforms, derivative product platforms, and standard interface platforms. We use the term integration platform to describe the architectural environment created to facilitate the reuse required for designing and manufacturing SOC applications in a consumer-driven market. Reasonable projections estimate that in just a few years we will need to achieve a reuse factor of greater than 96 percent to meet the productivity objectives. To reach such a level without compromising effective use of the silicon, reuse will be facilitated by application-specific and manufacturing-technology focused platforms designed to create the virtual sockets into which the reusable virtual components (VC) plug. The concept of an integration platform applies broadly to the ideas presented in this book. This chapter provides a high-level overview. For detailed discussions on designing and using integration platforms, see Chapters 6 and 7.

Targeting Platforms to the Market

An integration platform includes hardware architecture, embedded software architecture, design methodologies (authoring and integration), design guidelines and modeling standards, VC characterization and support, and design verification (hardware/software, hardware prototype). Because an integration platform derives much of its productivity by focusing on a particular target application, it begins with a characterization of that target market (for example, set-top boxes, digital cameras, wireless cell phones). However, many of the structural elements of a platform are shared across application domains.

Target Markets					
2G Wireless Phones	3G Wireless Phones	Digital Cameras	SOHO Network	HDTV	VIP

Business Models	Product Goals	Differentiators
ASIC Semiconductor Manufacturing	High-Volume Silicon	TTM, Cost, Process Technology, IP
Systems Product Manufacturing	End-Product Market Share	Feature Set, TTM (including Derivatives), Cost, Packaging
ASSP Product (Fabless)	High-Volume Chips	TTM, CSIC Derivatives, Cost, TTV
IP (VC) Provider	Architectural Franchise	Quality, Compatibility, Unique Functionality, Support
Design Service Provider	High-Value Services	Application Knowledge Skills, Design Methods

Figure 3.1. Target Markets and Their Business Models

A design platform is developed to target a range of applications. The range is a complex function involving several factors. These include the characteristics of the target market, the business model for engaging that market, the differentiation space the platform designer provides to the integrator, the time frame over which the platform will be used, and the process technologies on which it is implemented. Although most of these factors are well understood or explained in this book, it is worth noting how the application and motivation for a platform design varies depending on the business model used. The first row in Figure 3.1 provides a partial list of target markets. The columns identify some of the different business models that are involved in developing and manufacturing an SOC device.

The Role of Business Models

A platform's purpose and utility varies considerably when the perspective of a business model is taken into account.

ASIC Manufacturing

The primary goal of an application specific integrated circuit (ASIC) semiconductor vendor is the efficient production of silicon at a high level of its manufacturing capacity. An integration platform is a vehicle for the ASIC vendor to collect together in a prestaged form an environment that provides a significant differentiator beyond the traditional cost, technology, libraries, and design services. As the platform is reused across the family of customers in the target application market, the ASIC vendor sees direct benefits in terms of better yield in the manufacturing process, greater leverage of the VC content investment, a more consistent ASIC design handoff, and better time-to-market (TTM) and time-to-volume (TTV) numbers. Because the SOC consumer market is where rapid growth in high volume ASICs is projected, the ASIC providers were among the first to invest in platform-based methodologies.

System Product Manufacturing

The developer and manufacturer of the end product (that is, the cell phone, digital camera, set-top box) is a system company for which the SOC is a component in the final product. Because the SOC represents an increasing percentage of the product cost, value, and potential for derivative products, the system product designer has to consider the integration platform as an essential element in the product's life cycle. At a minimum, the system product designer uses a platform-like approach to rapidly spin design derivatives as required by the marketplace.

For instance, in the case of a digital camera, the development of an initial product might take a year, but it is then followed by a series of derivative product variations that must be to market on a much shorter development cycle (for example, two to four months or less). Failing to consider this in the initial design can significantly limit market share. Concurrently, the product cost is often forced to track a price erosion curve as competitors introduce newer products on more aggressive technologies. The systems designer of a consumer product uses a platform to respond to the market, to feature unique differentiating intellectual property (IP), to control costs, to diversify the supply chain, and to move to next-generation technology.

ASSP Providers

The application specific standard part (ASSP) provider designs IC chips (subsystem level) to be shipped as packaged parts and used in end products. Success in the market is often determined by the degree to which the ASSP is utilized in high-volume system products. To insure that this happens, the ASSP designer benefits significantly if the ASSP can be rapidly adapted to fit into high-volume applications. These adaptations, sometimes called CSICs (customer-specific integrated circuits), become increasingly more difficult as chip sizes and complexities grow. For the ASSP designer, the platform becomes the base for reuse to broaden the market opportunities for a part, while maintaining tight control over the design.

IP Providers

Independent IP or VC providers seek to have their components integrated into as many system and ASIC and CSIC platforms as possible. This means providing efficient interfaces to platform bus architectures as they emerge. It also means adapting and versioning the VCs to address disparate market requirements, such as low power, high performance, cost sensitivity, high reliability, or noise intolerance. The emergence of platforms for different application domains enables the VC provider to focus on a precise set of requirements and interfaces in pursuit of a particular market application.

Design Service Providers

Design service providers use platforms as vehicles for capturing their unique design methodologies and application domain differentiation. A platform represents the first codification of a new application domain and can be deployed through a design service that meets the TTM demands of the system product customer or positions the ASIC or ASSP provider to address new markets. Further, the platform is a structure around which an application-tailored design methodology can be fielded and reused. This methodology reuse is the basis for the design service provider achieving a differentiation in productivity and end-product design turnaround time, which can be leveraged across an application domain customer base, such as 3G wireless or small office/home office (SOHO) networks.

While the economic and competitive motivations among platform developers are varied, the fundamentals of platform design derive from a common source and apply generally. Basically, the notion of design platforms has developed by evolving the reuse paradigm into the system design context. Several new concepts emerge on the path from VC assembly to a more integration-centric, platform-design approach, some of which are the following:

- Adding software functionality along with software/hardware co-design and co-verification methods. This can include real-time operating systems (RTOS), drivers, algorithms, and so on.
- Investing in prestaging and verification of component combinations to be used as a fixed base for later incorporating into an SOC design. The prestaging combines the most critical design elements, such as processors, memories, analog/mixed signal (AMS) blocks, I/O structures, and bus/power/clock/test architectures.
- Codifying methods for assembling and verifying derivative products coming from the design platform. This creates a focus on the integration environment in terms of what can be done a priori and what appropriate limiting assumptions can be made after evaluating the target-application domain requirements.

Platform Levels

What emerges from this discussion of the purposes and uses of platforms is a collection of architectural layers, which make up the building blocks useful for constructing platforms. Figure 3.2 depicts a hierarchy of layers that reflects the development of integration platforms into fundamental technologies. Each layer builds upon those below it. As you move up the chart, platforms become progressively more application-focused, more productive, and more manufacturing-technology specific. While a variety of platforms could be defined within this

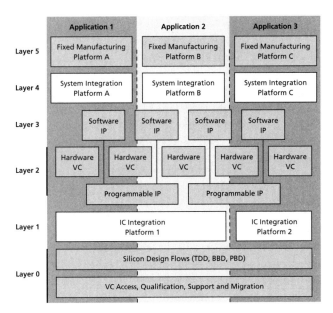

Figure 3.2. Integration Platform Layer Hierarchy

layering hierarchy, three platform levels are currently emerging as economically viable structures for investment.

The Foundation (Layer 0)

At the foundation, layer 0, the design methods for authoring blocks of IP (custom digital, AMS, standard cell digital), and integrating these blocks into a system chip, generally apply to all application domains. Additionally, the infrastructure environment for IP accessing, packaging, qualifying, supporting, and migrating to new technologies applies across all platforms. Before any effective design reuse environment can proceed, the foundation layer must be in place and accompanied by an appropriate cultural climate for acquiring and reusing design. However, this layer alone is insufficient for delivering significant productivity improvements resulting from design reuse.

Level 1: The IC Integration Platform

The IC integration platform, which spans layers 0 and 1, is the most application-general form of an integration platform that still delivers an observable improvement in reuse productivity. Like an operating system, it serves as the base upon which more application-focused platforms can be constructed. A typical level 1 platform consists of a set of programmable high-value hardware IP cores, which can be reused across a wide range of application sets, the RTOS for the processor(s), a system bus, a bridge to a peripheral bus, and an off-chip peripheral bus interface. Models for the cores are available and included in the provided methodology for hardware/software design verification. Typically, one or more representative peripherals are hung onto the peripheral bus, and the rudimentary operations among processors, memory, and peripheral(s) across the bus hierarchy are verified.

A level 1 integration platform can be applied to any application for which the selected processors, memories, and bus structure are appropriate. It is expected that during integration, application-specific IP will be integrated and adapted to the bus architecture. The power architecture, functional re-verification environment, test architecture, clocking, I/O, mixed signal, software architecture, and other integration elements are all defined during implementation, with changes made to IP blocks as needed.

The advantage of an IC integration platform lies in its flexibility and the reuse of the key infrastructural elements (processor, memories, and buses). The disadvantages lie in the adaptations required to use the platform to field a product, the need to re-verify the external VC blocks being integrated, and the wide range of variability in power, area, performance, and manufacturability characteristics in the finished product.

Figure 3.3 illustrates an IC integration platform that could be applied to many different applications in the wireless domain, from cellular phones to

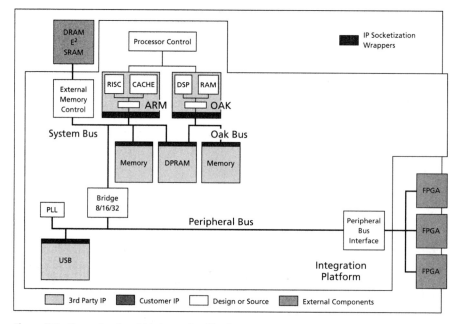

Figure 3.3. Example of an IC Integration Platform

SOHOs. Characterizing the processors, memories, buses, and I/O infrastructure provides a significant acceleration of any design in this space. However, to complete the product design still requires significant effort.

Level 2: The System Integration Platform

The system integration platform, which incorporates layers 0 to 4, provides a flexible environment for IP evaluation and exploration, system design, and rapid implementation in a variable die-size manufacturing structure. It is a means for end users to explore a variety of architectural alternatives and to integrate their own hardware and software differentiating algorithms. It also provides a very high degree of reuse in a predictable integration environment suitable for consumer-driven product development cycles. This platform is significantly more application-domain and process-technology specific than a level 1 platform or standard ASIC, with the expectation that over 95 percent of the design will be reused from the existing prestaged VC library without change. Primary product differentiation in this environment is achieved through unique product designer hardware blocks (analog or digital), software algorithms, and TTM. Manufacturing cost can be improved over a gate-equivalent ASIC or a level 1 platform by taking advantage of the fact that over 95 percent of the design and the key interconnect structures have been pre-characterized in silicon. Some of the key characteristics of the system integration platform are:

- Application-domain targeted
- Integration architecture specification (bus, power, clock, test, I/O architectures)
- Substrate isolation for mixed signal, required IP blocks, block constraints
- Full VC portfolio (prestaged, pre-characterized, pre-verified)
- Proven, scripted Virtual Socket Interface (VSI)-compliant VC authoring and integration methods
- Design guides for block authoring (register-transfer level, AMS, design for test, design for manufacturing)
- Verification environment and models (might include a prototype)
- Prototype characterization of integration architecture
- Embedded software support architecture (RTOS, debuggers, co-verification links/models, compilers, device drivers for peripherals, prototype emulation system for high performance software/hardware simulation)

Figure 3.4 shows an example of a system integration platform that addresses the DECT (Digital European Cordless Telephone) wireless market. This platform builds on the IC integration platform in Figure 3.3 by adding VCs and DECT-specific design content. It also includes a design infrastructure that is unique to the DECT application, such as AMS VC authoring noise management directives, manufacturing test techniques for the analog blocks,

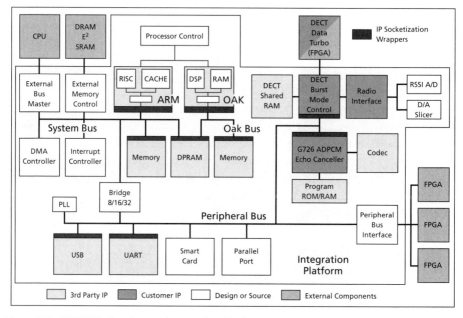

Figure 3.4. DECT Wireless System Integration Platform

additional bus extensions, external processor interfaces, device drivers for smart cards, and power management design techniques. This platform can still be customized by adding more peripherals, unique IP in areas such as the radio interface, cache size adjustments, software IP in the processors, and a host of other potential differentiators. Furthermore, this platform can be delivered as a predefined hardware prototype suitable for configuration with external field programmable gate arrays (FPGA) into an early application software development platform.

Level 3: The Manufacturing Integration Environment

The level 3 platform, which incorporates layers 0 to 5, represents a less flexible but more cost-effective and TTM-efficient vehicle for getting large SOC designs to market. It is very application-specific and most suitable for product markets and cycles where effective differentiation is achieved through cost, software, programmable hardware, or memory size adjustments. The manufacturing integration platform provides the user with a fixed die, fixed hardware development environment. The entire chip has been implemented, except for the final programming of the FPGA blocks and the embedded software in the processors. Some additional variability is achieved by using alternative implementations that include a different complement of memory and

FPGA sizes. The chip is delivered as its own prototype. Given the nature of many SOC products to be pin- or pad-limited, a wide set of variants on this platform style can be conceived, including ones as simple as containing just a processor, memories, and a large FPGA space all on a fixed die. However, without a high percentage of VCs already on board, implemented and characterized in an optimized fashion, and targeted at a specific application domain, the simpler versions will often fail to achieve competitive system chip cost, power, or performance objectives.

Perhaps the most significant advantage of this style of platform is seen in terms of manufacturing cost. The same silicon master is applied in multiple customer engagements, thus allowing significant manufacturing tuning for yield and cost.

There are also more complex variants to this level, such as reconfigurable computing engines with appropriately parameterized memories and I/O interfaces. Another unique approach is based on a deconfiguration technique, which offers a complete set of hardware already pre-configured and characterized, that enables the integrator to de-select chip elements and then program a new design into either software or FPGA. The chip vendor has the option of shipping the part in the original die (in the case of prototypes and early TTM small volumes), or shrinking the part for high-volume orders.

Using this integration platform methodology, we will now look at the systems-level entry point for SOC design—function architecture co-design.

4

Function-Architecture Co-Design

In this chapter, we take a "systems" approach to SOC design, using an integration platform methodology. This systems approach, called function-architecture co-design,[1] is based on an emerging class of co-design methodologies that extend beyond hardware-software co-design.[2] Function-architecture co-design begins at a level of design abstraction above that normally used in integrated circuit (IC) design, which today starts at register-transfer level (RTL) on the hardware side, and the C-code level on the software side. Instead, function-architecture co-design begins with a purely functional model of the desired product behavior, and abstract models of system architecture suitable for performance evaluation.

This chapter addresses why the function–architecture co–design approach is important for SOC design in light of the inadequacies of today's methodology. It also looks at the integration platform concept, the reuse of virtual components (VC) at a system level, and designing derivative products rapidly.

In terms of platform-based design (PBD), this chapter covers the tasks and areas shaded in Figure 4.1.

Changing to a Systems Approach

Adopting a systems approach to SOC design brings up many key questions and issues, such as:

- What is the function–architecture co–design approach to SOC design?
- Why do I need to change the current way of doing SOC design?

---

1. J. Rowson and A. Sangiovanni-Vincentelli, "Felix Initiative Pursues New Co-design Methodology," *Electronic Engineering Times,* June 15, 1998, pp. 50, 51, 74.

2. G. Martin, "HW-SW Co-Design: A Perspective," *EDA Vision*, vol. 1, no. 1, October 1997, www.dacafe.com/EDAVision/Issue1/EDAVision.1-3a.html.

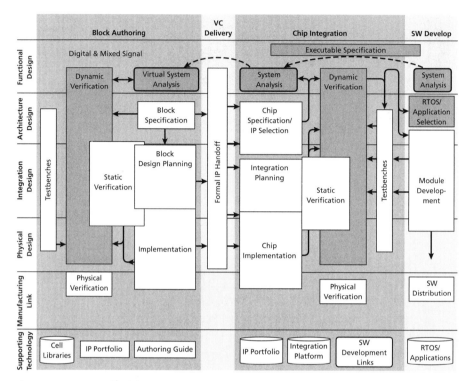

Figure 4.1. PBD Methodology: Function-Architecture Co-design Tasks

- Why can't I just evolve the way I do things today?
- How do I break through the "silicon ceiling"?
- What is the essence of this new methodology?
- How do I model the intended product behavior?
- How do I choose an appropriate SOC integration platform for my product?
- How do I model VCs to enable system-level design trade-offs?
- How do I improve the reusability of software VCs?
- How do I partition functions between hardware and software?
- How do I determine the correct on-chip communications and add architectural detail?
- How do I know that my system has the right performance?
- How do I choose the architectural components?
- How do I decide on processors without implementing a lot of software?
- How can I be sure that the communication and busing architecture is adequate?
- How do I reuse information and decisions made at the system level?
- How do I model integration platforms at the system level?
- How do I quickly design a derivative SOC device from an integration platform?

This chapter addresses each of these questions in the following sections.

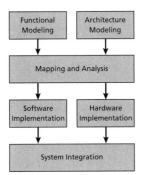

Figure 4.2. Phases of Function-Architecture Co-Design

Function-Architecture Co-Design

Figure 4.2 illustrates the main phases of function–architecture co-design when applied to embedded hardware–software systems.

Functional Modeling

In this phase the product requirements are established, and a verified specification of the system's function or behavior is produced. The specification can be executable, and it can also link to an executable specification of the environment within which the system will be embedded as a product. For example, a cellular handset might use a verification environment for a wireless standard such as GSM or IS-95 (CDMA). Functional exploration includes function and algorithm design and verification, probably simulation-based. The environmental executable specification can be considered a virtual testbench that can be applied to system-level design, as well as provide a verification testbench for implementing phases of the SOC design later.

Architecture Modeling

Once a functional specification for a product has been developed, a candidate architecture or family of architectures on which the system functionality will be realized is defined. Hardware/software architectures include a variety of components, such as microprocessors and microcontrollers, digital signal processors (DSP), buses, memory components, peripherals, real-time operating systems (RTOS), and dedicated hardware processing units (for example, MPEG audio and video decoders). Such components can be reused VC blocks, whether in-house, third-party, or yet-to-be-designed blocks. The system functional specification is decomposed and mapped onto the architectural blocks. Possibilities for component blocks include reconfigurable hardware blocks that are one-time, dynamic, with long reconfiguration latency, or dynamic with a short enough reconfiguration latency for the block to offer multiple function modes with considerable area efficiency.

Mapping and Analysis

This process maps, or partitions, the functional model onto the architecture model by assigning every function to a specific hardware or software resource: for hardware, as a dedicated hardware block (or one mode of a dedicated hardware block); for software, as a task running on a general or specialized processor. Embedded systems that contain several processors offer several choices for mapping particular software functions. They require at minimum a basic task scheduler, up to a complete RTOS, to mediate access to each processor resource. Although manual mapping can adequately deal with many systems, research into automated mapping algorithms might lead to methods that will be key to finding optimal mappings for the very complex embedded systems of the future.

After mapping, various kinds of performance analysis are possible, which might lead to experiments involving alternative architectures or alternative choices of VCs before an optimal architecture and mapping are found. A certain amount of architectural refinement can also be carried out prior to proceeding to the implementation phases.

Software and Hardware Implementation

This phase involves designing new hardware blocks, integrating reused hardware VC blocks, and developing software. Typical IC design today begins at this level of abstraction, often called the "RTL-C" level.

System Integration

With developed software and hardware, at least in prototype form, the complete system can be assembled for lab trials. Product integration might include emulators or rapid prototypes for hardware functions.

The SOC Design Process Today

Today, most system-level design of embedded SOC devices is based on giving a written specification for a product to an architectural guru, who then carries out a manual partitioning into chips or chipsets, writes a preliminary specification for the devices, and throws it to a chip development team. The team starts with RTL-coding and implements the chip using existing design flows, both logical (synthesis-based) and physical (floor planning-based). Behavioral modeling, often using C/C++, algorithm tools, and behavioral VHDL, is employed to some extent, with a limited degree of success, to work out the basic structure and partitioning of the system. Generally, however, such models are not shared between system architects and implementation teams.

In parallel, and often with poor or non-existent communication between the development teams, a software team develops code using their version of the specification. At integration time, hardware and software are brought together and iterated until they pass a verification test suite that is usually not comprehensive enough to guarantee type approval or product acceptance. This product development flow makes it a challenge to meet time-to-market (TTM) requirements.

Changing the Approach to SOC Design

Today's methodologies are largely geared toward authoring blocks on a low level subsystem basis, not integrating VCs into full SOCs. System-chip architectures captured at the RTL-C level are hard to reuse and evolve. At RTL, architectures must be fully articulated or elaborated, with all signals instantiated on all blocks, all block pins defined, and a full on-chip clocking and test scheme defined. Since architectural designs at RTL have completely defined communications mechanisms, it is difficult and time-consuming to change the on-chip control structures and communications mechanisms between blocks. It is also difficult to substitute VCs. Dropping in a new microcontroller core requires ripping up and re-routing to link the block to the communications structure.

In addition, designs captured at RTL mix both behavioral and architectural design together. Often the only model of VC function is the synthesizable RTL code that represents the implementation of the function. Similarly, the only model of a software function might be the C or assembly language implementation of the function. This intertwining of behavior and architectural components makes it difficult to evolve the behavior of the design and its architectural implementation separately. If a design needs to conform to a particular standard that is evolving, or needs to be modified to conform to the next generation of a standard, the RTL-C level design is a clumsy and difficult representation to work with.

Verification of embedded hardware-software designs at RTL is difficult, which is further compounded by having embedded products with a significant software component. At RTL, co-verification with today's ad hoc and emerging commercial tools is slow.[3] Complete system application behavior in a hardware description language (HDL)/C simulation environment cannot be

---

3. ibid.; and J. Rowson, "Virtual Prototyping," *CICC 1997*, May 1997, pp. 89-94.

verified. Co-simulation tools and strategies are still immature, constrained, and very slow (compared to system-level needs and algorithmic simulation at the system level). Rapid prototyping mechanisms, such as field programmable gate array (FPGA) and hard core-based, provide verification alternatives, but they generally require more complete software implementation or a concentration on specific algorithms.

During such co-simulation, if major application problems are found, a time-consuming and tedious redesign process is required to repair the design. Repartitioning is very difficult, since it causes the communications infrastructure to be redesigned. Substituting better programmable hardware VCs (new processors, controllers) or custom hardware accelerators for part of the software implementation requires significant changes to application software.

Why Today's Methods Do Not Work

As today's RTL tools and methodologies evolve, we will see more up-front system and chip design planning, better forward prediction of physical effects of layout (so that these effects can be incorporated into up-front design planning), and more robust hardware-software co-verification.

RTL, top-down floor planning is emerging. RTL floor planning offers better control over the physical effects determined during the synthesis process, and enables the number of iterations required to converge on a feasible design to be reduced. However, not all VC blocks will be reused at RTL. Reusing hard (essentially, laid out in an IC process) and firm (cell-level netlists) blocks will increase, and vendors of large complex programmable VC cores might prefer distributing these formats to their customers rather than synthesizable, soft RTL code.

VC block authoring is also becoming a better defined process as the Virtual Socket Interface (VSI) Alliance develops a VC block interchange standard covering the soft, firm, and hard domains from RTL down through physical layout.

However, such an evolution of methodology does not change the fact that architectures will still be hard to reuse and evolve. It will still be difficult to explore VC block alternatives, especially programmable ones. Verifying that an architecture and design work will still pose considerable difficulties. The behavior and architectural implementation for a design will still be intertwined and difficult to evolve separately.

This evolution also does not account for changing deep submicron (DSM) process technologies. Migrating design architectures to a new DSM-technology level requires porting the hard blocks to the new technology, which might not scale adequately for the required new applications for a derivative design, mapping firm blocks to new cell libraries optimized for the new process, and resynthesizing and re-characterizing soft blocks. The ported architecture and design might not meet the new derivative performance, cost, and power requirements, thereby requiring a significant redesign effort. DSM process migration for an integration platform might be better accomplished by taking a systems approach to evolving the process and making trade-offs and repartitioning decisions.

Adopting a New Methodology

Providing solutions for the limitations in today's methodology and tools requires moving away from concentrating on chip-level design at RTL-C level (referred to as the "silicon ceiling"),[4] that is, shifting from a methodology emphasizing VC authoring to one that emphasizes VC integration.

Breaking through the Silicon Ceiling

Breaking through the silicon ceiling requires higher levels of design abstraction in three key areas: architectures, models, and design exploration and verification.

Architectural abstractions must be easy to capture, evolve, and change, which means removing details that are not necessary for first- and second-order architectural exploration and evaluation. Abstract architectures are ideally suited to describing SOC integration platforms.

Architectural and VC choices cannot be explored with detailed cycle- and pin-accurate simulation models, because they are too slow to execute and too difficult to manipulate during design exploration. Articulated HDL-based signal and event-driven simulation, whether just used for hardware validation or as part of hardware-software co-verification, is also too slow to validate system-level behavior for embedded system-chip designs, or to explore architectural and VC alternatives. Instead, the appropriate abstraction level is to use performance analysis techniques to make first- and second-order architectural trade-offs. This matches the shift to a methodology centered on VC integration.

In addition, it is important to have a methodology that allows the system behavior to be repackaged or exported in an executable form to the lower levels of the design implementation process. This supports VC authoring (it can be used to verify new VCs that are part of the overall SOC device in the context of the intended system function as captured in an executable verification model), VC integration, and detailed design.

---

4. G. Martin, "Design Methodologies for System Level IP," *Proceedings of Design Automation and Test in Europe*, February 1998, pp. 286-289.

Breaking through the silicon ceiling also requires that RTL designers adopt a methodology and tool technology that supports:

- Easy, rapid, low-risk architecture derivative design based on standard integration platforms
- Reduced risk for developing complex product designs: system designers must be able to develop key concepts at an abstract level and have them rapidly implemented with little risk of architectural infeasibility
- Abstract models that are easy to generate
- First-order trade-offs and architectural and VC evaluations above RTL
- Hardware-software co-design and co-verification at a higher level
- On-chip communications schemes that are easy to create and modify, and the ability to hook up and change the interface of VC blocks to those mechanisms
- Linkages to hardware and software implementation flows
- Methods for parallel evolution of platforms as a function of process technology evolution to ensure that the latest technology is available

The Essence of This New Methodology

To maximize reuse and to take advantage of derivative designs, we need a new methodology to do the following (see Figure 4.3):

- Capture and iterate heterogeneous system behavior, both dataflow and control
- Compose behaviors by linking them with discrete event semantics
- Capture a minimal or relaxed product architecture
- Manually map the behavior to the architecture
- Annotate the behavior with architectural performance effects in terms of speed, power, and cost, using architectural estimation models
- Carry out a performance analysis of the behavior on the architecture and iterate
- Refine the architecture to an implementable hardware-software μ-architecture that can be passed to hardware and software implementation flows

Putting the New Methodology into Practice

The following section provides a methodology for creating derivative designs.

Modeling the Intended Product Behavior

The system designer captures and verifies the functional behavior of the entire system at a pure behavioral level. This is based heavily on reusing behavioral libraries and algorithmic fragments, importing language-based models of behav-

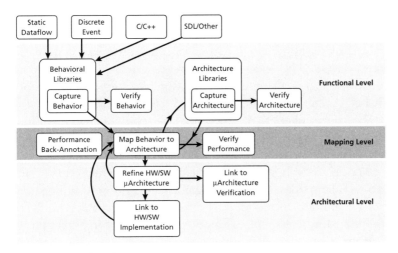

Figure 4.3. Design Flows

ior, and creating co-design finite state machine (CFSM) models.[5] The verification occurs within an implementation-independent environment; that is, no architectural effects are incorporated into the functional behavior at this point.

A target architecture is devised and captured, which is based on reusing architectural VC elements, such as DSP cores, microcontrollers, buses, and RTOSs. Architectures are captured in a relaxed mode, without wiring detail.

Choosing an Appropriate SOC Integration Platform

Target architectures are the mechanism for describing domain-specific integration platforms. Target architectures can be provided as integration platform definitions rather than starting from scratch. In this case, the designer starts with a platform and a target application and explores the space of architectural modifications that are possible using the platform's VC portfolio.

Modeling VCs to Enable System-Level Design Trade-offs

First-order architectural trade-offs do not need fully articulated architectures to be captured. A "relaxed" view of the system-chip architecture is sufficient. In this relaxed view, VC function blocks are instantiated with simple connections to abstract views of communications mechanisms. At the highest level of

5. F. Balarin, M. Chiodo, P. Giusto, H. Hsieh, A. Jurecska, L. Lavagno, C. Passerone, A. Sangiovanni-Vincentelli, E. Sentovich, K. Suzuki, and B. Tabbara, *Hardware-Software Co-Design of Embedded Systems*, Kluwer Academic Publishers, Dordrecht, The Netherlands, 1997.

abstraction, communications can be described as moving frames, packets, or tokens between function blocks over channels. Below that, communications abstraction is seen as a series of basic bus-transactions or software communications methods.[6]

The VC function blocks can be classed into several categories: processors (control-dominated or signal processing-dominated), custom function blocks (for example, MPEG decoders, filter blocks), memories, peripheral controllers, buses, etc.[7] These VC function blocks process tokens, frames, packets, or step through control and computational sequences under software control. The basic system operation can be described by how fast blocks process tokens or run software, and how blocks transfer tokens to each other over communications mechanisms.

The abstract VC models should be a combination of architectural delay equations, appropriate for the class of VC block, and resource contention models for shared resources, such as buses and processors.

Improving the Reusability of Software VCs

Software code is reusable if it can be easily retargeted. There are two kinds of software VCs:

- Close-to-hardware, consisting of RTOSs, drivers, and hardware-dependent code that is optimized for particular hardware platforms. This software is often written in assembly code and inherently hard to retarget.
- Hardware-independent, usually written in C with adequate performance when kept as hardware-independent. Retargeting requires an assurance that the software will perform adequately on new target hardware. Techniques currently exist (based on extensions to the software estimation work in POLIS)[8] that enable this assurance to be derived by estimating software performance automatically on target hardware.

To ensure software reusability, VC developers should write hardware-portable code using APIs, compiler directives, and switches to invoke various hardware-specific functions or code.

Partitioning Functions between Hardware and Software

Behavioral functions and communications arcs are manually mapped to the architectural resources, and the system is evaluated using a performance analysis of

6. J. Rowson and A. Sangiovanni-Vincentelli, "Interface-based Design," *Proceedings of the Design Automation Conference*, 1997, pp. 178-183.

7. G. Martin, "Moving IP to the System Level: What Will it Take?," *Proceedings of the Embedded Systems Conference*, 1998, Volume 4, pp. 243-256.

8. F. Balarin, H. Hsieh, A. Jurecska, L. Lavagno, and A. Sangiovanni-Vincentelli, "Formal Verification of Embedded Systems Based on CFSM Networks," *Design Automation Conference*, 1996.

speed, power, and cost. Pre-existing architectures and mappings can provide start-ing points for this phase. The process of mapping establishes a set of relationships between the application behaviors and the architecture on which that behavior will be realized. When several behavioral blocks are mapped onto a programma-ble VC block, such as a microprocessor, controller, or DSP, it is assumed that these behaviors are intended to be implemented as software tasks running on the processor. This might involve at minimum a simple scheduler, or a commercial or proprietary RTOS. When a behavioral block is mapped onto a dedicated hard-ware unit on a one-for-one basis, we can assume that the hardware block imple-ments that behavior in hardware (see Figure 4.4 as an example of this).

Where there is no existing architectural resource on which to map a behav-ioral function, or where available resources from VC libraries are inadequate in performance, cost, power consumption, and so on, the system behavioral requirements and constraints constitute a specification for a new architectural resource. This can then be passed, along with the behavioral model and the overall system model, to an implementation team to design new VCs accord-ing to this specification. Alternatively, trade-offs can be made as far back as the product specification to reach the appropriate TTM—the need for the new function is traded with the extra design time required.

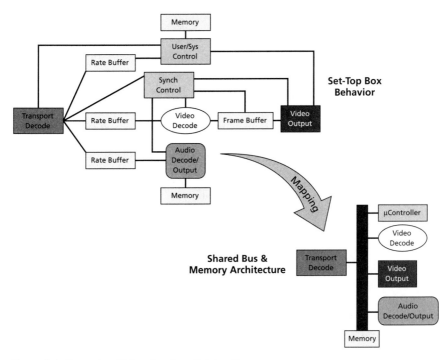

Figure 4.4. Example of Mapping for a Set-top Box

Determining the Correct On-Chip Communications

When a communications arc is mapped onto a communications resource on a one-for-one basis, usually there is no contention for the resource. When several communications arcs are mapped onto a communications resource, the resource is shared and, thus, contended for. Communications mapping starts at a very high level of token-based communications abstraction, which is later refined to add detail.

The architecture, behavior, and mapping are iterated and analyzed until the most optimal architecture is established. The target architecture is refined into a detailed micro-architecture that is implementable in both hardware and software domains. For example, memories are mapped onto actual implementations, detailed communications interface blocks are chosen using generic bus transactions, and glue-control hardware is defined, possibly as CFSMs. The refined target architecture is passed to subsequent processes for hardware and software implementation.

For new VC blocks, or for existing VC libraries that do not meet system requirements for some behavioral functions, the results of detailed implementation can be abstracted and back-annotated into the system model to ensure that block implementations still work within the overall system application behavior. Any changes that could affect the behavior are verified within the original verification environment or executable system model.

Determining the Right Performance

After mapping, the architecture's performance in running the application must be analyzed or compared with alternative architectures. In this methodology, a performance analysis simulation is carried out by re-running the behavioral simulation annotated with information derived from architectural estimation models (delay equations), which represent the performance effects of the architecture under study.

The netlists representing the behavior and architecture, and the mapping between them, are combined with delay equations and resource models extracted from architectural component libraries to regenerate a behavioral composition netlist, with annotations representing delay equations and resources.

The behavioral simulation is then re-run in native mode, processing testbench traffic, while keeping track of the performance effects of the architecture (the delay equations and resource contentions) via instrumentation on the simulation. This simulation, which keeps track of cycle counts, is called cycle-approximate functional simulation.

The output of this instrumented simulation can then be analyzed via a number of visualization tools. Architectural changes can be evaluated to find the optimal architecture for the application.

Choosing the Architectural Components

Each architectural component has an appropriate mechanism for creating delay equations and for evaluating performance effects. These components become part of the VC portfolio provided with the integration platform. The portfolio of VCs is used to explore the mapping space. Architectural resources can be divided into several different classes:

- *Estimatable processors* Software tasks mapped to estimatable processors, such as microcontrollers, RISC, and CISC, wait to get access via the RTOS scheduling policy, estimate performance based on characterized estimation function coefficients, and then release the processor. This is especially suited for control-oriented code, which has high input-data dependency. The estimation procedure is further described in the next section.

- *Non-estimatable processors* These processors are modeled using a series of DSP kernel functions that are pre-characterized on the DSP through analysis or through running assembly or C code for the kernels on the processor, which are then used to derive an equation-based model. During performance simulation, the software functions running on the DSP are mapped into an appropriate set of kernel equations. Contention for the DSP is also modeled via an RTOS or simple scheduler model. This is especially suited for dataflow-oriented code that has predictable latency and relatively high data independence in the flow or processing.

- *Software tasks* If multiple behavioral blocks are mapped to a single task, they will be statically scheduled, and communications within the task avoid RTOS overhead.

- *Buses* Buses are modeled through a set of basic bus transactions (for example, write, burst-write, read, burst-read), which are characterized individually via delay equations. In addition, contention for the bus is modeled as a resource. Behavioral communications arcs that are mapped to specific sets of bus transactions also have a set of transactions they can perform.

- *Memory* Simple models for memories represent wait states as delays. Software will both execute out of memory and use it for data store and access. Memory hierarchy via cache mechanisms can be modeled stochastically or via more sophisticated resource models. Software estimation techniques factor necessary memory accesses into them. Both Harvard and non-Harvard memory architectures can be modeled.

- *Hardware* Existing blocks have performance delay equations for processing tokens, frames, packets; new custom hardware blocks have either constraints or back-annotated delay numbers derived from implementation. These delay equations can represent dedicated function block performance, whether realized through custom digital design, application-specific IC (ASIC) styles (for example, standard cell synthesis, placement, and routing), embedded FPGA blocks, or even the use of newer approaches, such as dynamically reconfigurable hardware.

* *RTOS* Each RTOS or variant, whether commercial or proprietary, should have a dynamic resource model for scheduling policy, interrupt service, and context-switching latencies.

Using Processors without Implementing a Lot of Software

The technique used for estimating the performance of software running on a target processor or microcontroller core is based on two key steps. First, the processor is characterized once, and a table of coefficients is created, as shown in Table 4.1, and placed in the library. These coefficients give cycle counts for a basic set of generic or atomic operators, which are common across all processors and controllers for a class of applications. The generic or atomic operators map into actual processor instructions.

Next, the C code is analyzed and decomposed into the generic atomic operators, which are annotated with delay coefficients from the table. Processor register resources are used to estimate variable storage in registers versus memory accesses. During performance analysis simulation, the actual software code is run natively on the workstation and accumulates delay information based on counting the cycles that would have been consumed on the target processor. Statistical and scaling techniques model cache and pipelining effects.

Table 4.1. Example of a Coefficient Table

Operator	Description	Delay in Cycles
LD	Load from data memory	3
LI	Load from instruction memory	1
ST	Store to data memory	2
RR	Register-to-register move	1
OP	Simple ALU operation	1
OX1, OX2	Complex ALU Operation	17, 39
TS	Test and branch	1
BR	Unconditional branch	3
BS	Branch to subroutine	19
RT	Return from subroutine	18

Some code might have hybrid characteristics: data dependencies in some portions, and data independence in others. Several hybrid estimation schemes can be used, depending on the granularity of the mix of control and dataflow. For example, either of these methods, or a combination of them, can be used:

- If control and dataflow code is mixed within the tasks at a fine-grained level, the control software estimation method can be used for the major control flow. If the code then calls pre-characterized DSP kernel functions, a statically or parametrically driven model for the kernel function latency can be used.
- If control and dataflow processing exhibit task-wise granularity, one RTOS scheduling model can be used to mediate access to the processor, but either the DSP kernel function modeling method, or the control software estimation method, can be used on each task depending on its dominant type.

Communication and Busing Architecture

Communication refinement is the process of mapping communication arcs in the behavioral hierarchy to architectural resources and decomposing the resulting communication blocks down to the pin-accurate level. Arcs connecting behavioral blocks mapped to the same software processor are only decomposed down to the RTOS interface level. If, within the architecture specification, standard hardware components or a standard RTOS are selected, these selections constrain the decomposition process on the behavioral side to match the actual interfaces within the architecture. This approach is known as interface-based design.[9]

At this level, the mapped behavior is extended to model the communication mechanisms within the system architecture. For two communicating behavioral blocks mapped to hardware (hardware to hardware), the modeling is done at the generic bus transaction level. For example, the user sees transactions such as write(501), read(), burst_write(53, . . .), irq(1,5). The token types transmitted are those directly supported by the hardware bus. Transactions are modeled as atomic units and have no signals or internal timing structure. The actual signals of the bus are not modeled, nor things like bi-directional ports or tristate drivers. Shared resources within the implementation (processors, busses, etc.) are modeled abstractly via a shared-resource model and are instantiated by the performance simulation interpretation of the delay equations.

Software to hardware or hardware to software communication is also modeled at the same bus transaction level. The refinement on the software side reflects the chosen RTOS's device drivers. Modeling software to software communication at this level also occurs when the two software behaviors are mapped to different processors.

9. Alberto L. Sangiovanni-Vincentelli, Patrick C. McGeer, and Alexander Saldanha, "Verification of Electronic Systems," *Proceedings of the Design Automation Conference*, June 1996, pp. 106-111; and Rowson and Sangiovanni-Vincentelli, "Interface-based Design."

For software to software communication, if the two software behaviors are mapped to the same processor, the communication interface is modeled with the RTOS, since it is not necessary to model at a lower level of abstraction. The transaction types in this case might be wait(), read(), lock(), unlock(), emit().

This refined behavior modeling reflects the lowest communication abstraction level that can be modeled and simulated. Every mapped communication arc must be modeled at this level before it can be exported from the methodology. For this reason, all communication resources in the system architecture (buses, RTOSs, interfaces, etc.) *must* provide a model at this level of abstraction.

Additional abstraction levels are possible between the mapped behavior and the refined behavior levels. Modeling at these levels is optional, and models for these levels might or might not be available.

Performance analysis of communications traffic at the system level, down to generic bus transactions, provides useful block to block latency and bandwidth requirements that can be used later to determine the detailed bus structures and arbitration schemes when implementing the platform and derivative designs. After implementation, more accurate bus latency and throughput characteristics can be back-annotated into the system-level models. This can be used to explore architectural alternatives and to improve the fidelity of the platform system models for subsequent derivatives.

Reusing Decisions Made at the System Level

This methodology produces an implementable hardware and software description. The hardware description is passed to an RTL floor planner and to a cycle and pin-accurate HDL verification environment. The hardware description consists of:

- A top-level HDL file with references to all the VC blocks, wired together with full pin-accurate wiring (for example, all signals referenced), including I/O pads, test buses, self-test structures, and parameters for parameterized VC blocks.
- Synthesizable RTL HDL blocks (invoked from the top level) where communications structures have been chosen in the refinement process, or in the software code that implements the communications structure (for example, a bus interface), along with appropriate performance constraints.
- An assumption check testbench that helps validate at the cycle-accurate level the assumptions made at the performance analysis level, which include each library delay equation, the tool's delay calculations where library equations do not exist, each communication mechanism between blocks, the RTOS and/or bus arbitration operation, a function behavior and performance (marked by the user) when the function operation is highly data dependent, and the memory size (data and code).

The software description consists of:

- Each software implementation described as a memory image, with specific physical memory location, size, and load address.
- A memory map with no overlaps, including a fully defined interrupt vector and direct memory access trigger table.
- A skeleton of the overall software structure, including initialization code where available, and calls to major RTOS setup routines.

As mentioned above, the communications bandwidth and latency requirements can be directly passed to a detailed design of on-chip buses.

Using the Integration Platform Approach

The integration platform approach enables a chip architecture to be reused as a whole if it is supported by an efficient system-level design methodology, such as function-architecture co-design.

Modeling Integration Platforms at the System Level

To model integration platforms at the system level, architectures must have the following characteristics:

- Simple to capture and modify, that is in a relaxed form rather than a fully articulated form.
- Include rich libraries of architectural VC components from internal and third-party VC providers.
- Supported by central architecture groups and third-party suppliers who create architectural derivative product design kits containing reference architectures, VC block libraries, and sample applications.
- System control and communications that are easy to modify by using abstract communications descriptions and refinement mechanisms.
- Easy to export to implementers of architectural derivatives. It must be possible to link architectural design to real hardware and software implementation flows, so that design information captured at the architectural level is usable at subsequent design process stages.

Designing a Derivative SOC Device from an Integration Platform

In today's embedded consumer communications and multimedia products, original architectures created on a blank sheet are relatively rare. However, a base or platform architecture is often used to create a whole series or family of derivative products. Derivative designs can rely on the same basic processor cores and communications buses, but they can also be varied in the following ways:

- Change peripherals depending on the application
- Add optional accelerating hardware
- Move hardware design into software, relying on new, faster embedded processor cores or parallel architectures (for example, very long instruction word (VLIW) architectures)
- Limited VC block substitution (for example, moving to a new microcontroller core which, via subsetting, is instruction-set compatible with the old core)
- Significantly change software, tailoring a global product for particular markets or adding special user interface capabilities

Sometimes a derivative design is only possible if the whole integration platform on which it is based is moved to a new DSM process that provides greater performance, lower power, or greater integration possibilities.

Ideally, system designers would be supplied with an application-oriented architectural template toolkit, that is an integration platform, for constructing derivative SOC designs. This toolkit, which defines a virtual system design, would contain the following:

- A template architecture or architectural variants, including basic processing blocks, SOC on-chip and off-chip communications buses, basic peripherals, and control blocks.
- An application behavior, along with a verification environment or testbench.
- A "starter" mapping of behavior to architecture.
- Libraries of behavioral and architectural components that could be used to create a derivative architecture for a modified behavior.
- Composition and refinement rules and generators that would keep system designers in the feasible derivative space.

In this template toolkit, system designers might want to modify the application behavior for a particular design domain, for example, to incorporate a new standard. Behavioral libraries, which are regularly updated, should include new standards-driven blocks as standards emerge or evolve.

Also, the current template architecture might not meet the system constraints with the new behavior, and a new architectural VC component might need to be added. The architectural VC component libraries should be regularly updated to include new function blocks, new controllers, and so on.

The mapping of behavior to architecture should be modified to incorporate the new components and the performance analysis redone to validate system conformance.

Using the refinement rules and generators supplied by the central architecture/VC group, a new set of implementation deliverables can be generated and passed to an implementation group.

What's Next?

Several key factors are important for this new methodology to succeed:

- The availability of VCs that can be incorporated into integration platforms and used to construct derivative designs based on this system-level trade-off methodology. These VCs need to have the appropriate abstract system models. Existing and emerging VCs need to be modeled, and the models made available to a wide user community.
- Appropriate interface-based design models at all levels of the design hierarchy need to be used, since this approach promotes a modular development strategy where each architectural VC is developed, tested, verified, and pre-characterized independently.
- Organizations must develop internal structures to ensure effective VC management, information sharing, central or distributed VC databases to allow VCs to be found, and careful planning to avoid redundancy in VC development and to promote wide reuse.
- Careful VC database and design management so that the impact of VC revisions on past, current, and new designs can be carefully assessed prior to checking in the new versions. This will also help identify a VC development and procurement program for a central organization.

Moving beyond today's RTL-based methodology for system-chip design requires a new reuse-driven methodology and the provision of tools and technologies that support it. Function-architecture co-design provides solutions to taking a systems approach to SOC design.

Table 4.2. System Design Solutions to Key SOC Issues

Issue	Recommended Solution
What is the function-architecture co-design approach to SOC design?	Behavior-architecture co-design: orthogonal capture of system function, implementation architectures (integration platform), and explicit mapping between the two.
Why do I need to change the current way of doing SOC design?	Current approaches do not link system design and implementation. As a result, too often the system is designed on paper.
Why can't I just evolve the way I do things today?	You will still be stuck having to implement SOC systems at RTL-C level before finding out they won't work.
How do I break through the "silicon ceiling"?	Define abstractions of behavior and architectures, and architectural components.
What is the essence of this new methodology?	Maximize reuse and take advantage of derivative designs.

(Continued on next page.)

Table 4.2. (*Cont.*)

Issue	Recommended Solution
How do I model the intended product behavior?	Use behavioral models from many sources, including standards-based libraries, C/C++ models, CFSM models.
How do I choose an appropriate SOC integration platform for my product?	Map behavior to candidate platform architectures and carry out analysis.
How do I model VCs to enable system-level design trade-offs?	Use performance delay equations and software estimation methods. Eventually, will extend to power equations and cost models.
How do I improve the reusability of software VCs?	Careful structuring of software components into hardware-dependent and hardware-independent layers; use of APIs and retargetable high-level languages.
How do I partition functions between hardware and software?	Explicit mapping of function to hardware and software architectural components, with careful analysis of results.
How do I determine the correct on-chip communications and add tectural detail?	Mapping includes both hardware and software communications mechanisms, modeled first at archi-a token-based abstraction level.
How do I know that my system has the right performance?	Performance analysis of mapped function-architecture.
How do I choose the architectural components?	Use the portfolio of VCs provided with the integration platform and explore the mapping space. Processors are modeled with software estimation techniques for both control and dataflow code.
How do I decide on processors without implementing a lot of software?	Software estimation and use of kernel functions.
How can I be sure that the communication and busing architecture is adequate?	Refine communications to generic bus transactions with more accurate delays and resimulate system behavior.
How do I reuse information and decisions made at the system level?	Links to implementation: generate detail of hardware and software micro-architectures in imple-mentable form.
How do I model integration platforms at the systems level?	Create design kits of target application behaviors, architectures, VC portfolio models, and sample mappings. Use these kits to design derivatives.
How do I quickly design a derivative SOC device from an integration platform?	Use system-level design kits modeling platforms and allowing quick derivative construction.

5

Designing Communications Networks

Modern, complex electronic system designs are partitioned into subblocks or subsystems for various reasons: to manage the complexity; to divide the system into complete functions that can operate independently, thus simplifying interblock communications and allowing for parallel operation of subfunctions within the system; to minimize the interconnect or pins for each subblock for ease of block assembly; or to specify subsystems in a way that enables using standard blocks or previously designed subfunctions. Complex systems require the successive refinement of models from the very abstract algorithmic level down to a partitioned block-based architecture.

The partitioned system must then be reassembled with the appropriate communications network and information exchange protocols so that the overall system functionality and performance requirements can be met. The communications must also be modeled as a successive refinement, leading to a set of transactions. The function-architecture co-design methodology introduced in the previous chapter does this refinement. We will now discuss the implementation of bus architectures starting at the transaction level.

This chapter describes the fundamentals of bus architecture and techniques for analyzing the system-level communication of a design and applying it to the bus creation in a block-based chip-level design. It begins with definitions and descriptions of key communication components followed by design methodology. It includes a detailed discussion on adapting communications designs to a platform-based paradigm. Engineering trade-offs and a look at the future of on-chip communications is also addressed.

In terms of the platform-based design (PBD) methodology introduced earlier, this chapter discusses the tasks and areas shaded in Figure 5.1.

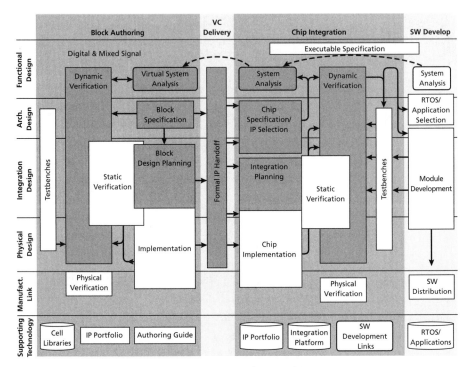

Figure 5.1. PBD Methodology: Implementation of Bus Architecture

Overview of System Chip Communications

The communications network provides for the sharing of information between functions and for the transmission of data and control information between individual functions and with the outside world. This communications network can be viewed as having both physical and logical entities. The physical view consists of a hierarchical network of elements, such as bus structures, ports, arbiters, bridges. The logical view contains a hierarchical set of information exchange protocols.

Before introducing a methodology for designing communications networks, this section defines and describes the important elements and concepts in system chip communications.

Communication Layers

Generally, system chip communications can be divided into hierarchical layers. The lowest layer includes the physical wires and drivers necessary to create the network. At this layer, the physical timing of information transfer is of key importance. At the next layer, the logical function of the communications network is defined. This level includes details on the protocol to transfer data between subcomponents or virtual components (VC). These first two layers deal with the specific implementation details. The top-most layer, or applications layer, describes interactions between components or VCs. This layer does not define how the data is transferred, only that it gets there. Therefore, a third layer is needed to bridge the lower implementation layers and the upper application layer. This third layer is the transaction layer. A transaction is a request or transfer of information issued by one system function to another over a communications network. In this layer, transactions are point-to-point transfers, without regard to error conditions or protocols.

The transaction layer is key to understanding and modeling VC to VC communications, because it is removed from the bus-specific details but is at a low enough level to transfer and receive data in a meaningful way from VCs. The transaction layer corresponds to the lowest level of communications refinement in systems design, as discussed in the previous chapter. The VC interface should be defined as close to the transaction level as possible.

The transaction layer can be further subdivided into two levels: the higher level dealing with abstract transactions between modules, and the lower one dealing with transactions closer to the hardware level. The higher level consists of reads or writes to logical addresses or devices. These reads and writes contain as much data as is appropriate to transfer, regardless of the natural limits of the physical implementation. A lower level transfer is generally limited by the implementation, and contains specific addressing information.

Table 5.1. System Chip Communications Layers

Layer	Name	Definition
4	Application	Deals with many interacting components. Contains conventions for memory maps and status or control information.
3	Transaction	Point-to-point transfers between VCs. Covers the range of possible options and responses (VC interface).
2	Bus Transfer	Protocols used to successfully transfer data between two components across a bus.
1	Physical	Deals with the physical wiring of the buses, drivers, and timing specific to process technology.

Buses

Buses are a way to communicate between blocks within a design. In the simplest form, buses are a group of point-to-point connections (wires) connecting multiple blocks together to allow the transfer of information between any of the connected blocks. Some blocks might require the transfer of information on every clock cycle, but most blocks within a system need information from other blocks only periodically. Buses reduce the amount of pins needed to communicate between many different units within the system, with little loss in performance.

To partition a system into subfunctions, logic must be added to each of the blocks to keep track of who gets to use the bus wires, when is the data for this block, when should the sender send the data, did the receiver get the data, etc. The bus also requires control signals and a protocol for communicating between the blocks.

There are many different ways to create a bus protocol. In the simplest case, one device controls the bus. All information or data flows through this device. It determines which function sends or receives data, and allows communications to occur one at a time. This approach requires relatively little logic, but does not use the bus wires efficiently and is not very flexible. Another approach is for all the communications information to be stored with the data in a packet. In this case, any block can send data to any other block at any time. This is much more flexible and uses the bus wires more efficiently, but requires a lot of logic at each block to determine when to send packets and decipher the packets being received. The former example is traditionally called a peripheral bus, and the latter is called a packet network.

Bus Components

The initiator, target, master, slave, and arbiter bus interface functions are types of communications processes between subfunctions or VCs. Bridges are used to communicate between buses.

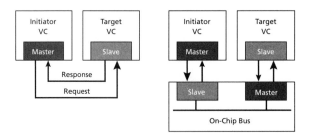

Figure 5.2. Bus Components

An initiator is a VC that initiates transactions. It defines the device and address it wishes to access, sends the request for the transaction to the bus, gets a grant from the bus arbiter, and responds to any error that might occur. A target is a VC that only responds to transaction requests; it never initiates requests or transactions.

A master and initiator are often interchangeable, but we are differentiating between an initiator as the component on the bus, and a master as an interface. A master then is the initiator side of the VC interface. Similarly, a slave is the target side of the VC interface.

An arbiter controls the access to the bus. All requests must be directed to the bus arbiter, which then arbitrates the sequence of access to the bus by the VCs.

A bridge connects two buses together. It acts like an initiator on one side of the bridge and a target on the other. Bridges can have intermediate storage to capture part of or the entire transfer of information before passing it on to the next bus. Bridges can also change from one size bus to another.

The arbiter decides to which initiator to grant the bus. This is important when multiple initiators exist on a bus. Each can initiate transfers of data, but if more than one wants to initiate a transfer, the arbiter decides which one gets the bus first. As shown in Figure 5.3, arbitration can be done in different ways. With a serial scheme, all the initiators on a bus have a strict priority; the one closest to the arbiter always gets the bus first, then the next, and so on down the line.

In the parallel approach, all of the initiators request to the arbiter in parallel, and, generally, it is first come, first serve, with some obvious implicit priority based on the structure of the logic internal to the arbiter. In polling, each initiator gets priority to use the bus in turn. For example, if one initiator gets priority on the first cycle, the next one gets priority on the next cycle, and so on until all devices have had a turn, at which time the cycle repeats itself. Some

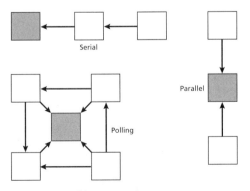

Figure 5.3. Arbiter Examples

versions of PCI have a two-tiered form of polling, where all high-priority devices are given a turn before the next lower-priority device gets a turn.

In most arbitration schemes, the devices must request the use of the bus. This means that a device that does not have priority to get the bus on this cycle could still get it if the devices of higher priority are not requesting the bus on that cycle.

Bus Hierarchy

Within a typical processor-based system, the hierarchy of buses goes from the highest performance, most timing critical to the lowest performance, least timing critical. At the top of this hierarchy is a processor bus, which connects the processor to its cache, memory management unit, and any tightly coupled co-processors. The basic instructions and data being processed by the CPU run across the bus. The bus affects CPU performance and is usually designed for the highest performance possible for the given implementation technology (IC process, library, etc.). Because the bus's configuration varies with each processor, it is not considered a candidate for a VC interface.

At the next level, all of the logically separate, high-performance blocks in the system, including the processor, are connected to a high-speed system bus. This bus is typically pipelined, with separate data and address. It usually has more than one initiator, and therefore contains some form of arbitration. The processor has a bridge between its internal bus and this system bus. The system bus usually contains the memory controller to access external memory for the processor and other blocks.

The lowest level of the hierarchy is the peripheral bus. Usually a bridge exists between the peripheral bus and the system bus. Typically, the only initiator on a peripheral bus is the bridge. Since the peripheral bus provides communications to the interfaces with functions that connect to the outside world, most of the devices on a peripheral bus are slow and generally only require 8-bit transfers. The peripheral bus is, therefore, simpler, slower, and smaller than a system bus. It is designed to save logic and eliminate the loading penalty of all the slow devices on the system bus.

Bus Attributes

When describing or specifying a bus, you must identify its latency, bandwidth, endian order, and whether it is pipelined.

Latency is the time it takes to execute a transaction across the bus. It has two components: the time it takes to access the bus, and the time it takes to transfer the data. The first is a function of the bus's protocol and utilization. The second is directly determined by the protocol of the bus and the size of the packet of data being transferred.

Bandwidth is the maximum capacity for data transfer as a function of time. Bus bandwidth is usually expressed in megabytes per second. The maximum bandwidth of a bus is the product of the clock frequency times the byte width of the bus. For example, a bus that is clocked at 100 megahertz and is 32-bits wide has a maximum bandwidth of 400 megabytes per second, that is 4 bytes per clock cycle times 100 million clock cycles per second.

The effective bandwidth is usually much less than the maximum possible bandwidth, because not every cycle can be used to transfer data. Typically, buses get too many collisions (multiple accesses at the same time) if they are utilized above about one-third of their maximum capacity. This effective bandwidth can be lower when every other cycle is used to transfer the address along with the data, as in buses where the same wires are used for the transfer of both data and address, and is higher with more deeply pipelined separate address and data buses.

Pipelining is the interleaved distribution of the protocol of the bus over multiple clock cycles. In Figure 5.4, the activity on the bus consists of successive requests to transfer data (a,b,c,d,e,f,g,h,i). First the master makes the request, then the arbiter grants the bus to the master. On the next cycle, the bus initiates the transfer to or from the target. After that, the target acknowledges the request and on the following cycle sends the data. Not all buses have as deep a pipeline as this. In many cases, the acknowledge occurs in the same cycle as the data. In other cases, the grant occurs in the same cycle as the request. While there are many variations to pipelining, they all serve to improve the bandwidth by eliminating the multiple dead cycles that occur with a non-pipelined bus.

The endian order determines how bytes are ordered in a word. Big endian orders byte 0 as highest; little endian orders byte 0 as lowest. Figure 5.5 shows the addressing structure.

The big endian scheme is appropriate when viewing the addressed text strings, because it proceeds from left to right, because the byte address within the word when added to the word address (in bytes) is equivalent to the actual byte address, as can be seen in Figure 5.5. For example, byte 2 of word 0 in the little endian example above would be byte 2, and byte 3 in word 1 would be byte 7 by byte addressing. This is the word size (4) times the word address plus

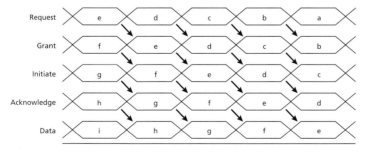

Figure 5.4. Pipelined Bus

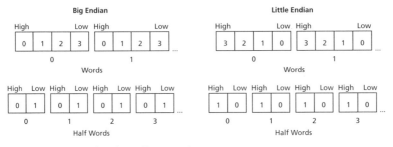

Figure 5.5. Big and Little Endian Words

the byte size (4 ★ 1 + 3 = 7). In the big endian case, the same byte is word 1, byte address 0.

Similarly, if the above examples contained the characters ABCDEFGH in the two successive words, the big endian example would be ABCD EFGH, where the little endian would be DCBA HGFE. If characters were the only type of data, matching byte addresses would be sufficient. When you mix the two addressing schemes for words or half words it is more of a problem, because the mapping is context dependent, as can be seen in the diagram above. A system chip might have some functions that require big endian and some that require little endian. When a VC that is designed for little endian requests data from a VC that is designed for big endian, translation from big endian to little endian is required. In this case, additional logic for conversion must be in the wrappers.

VSI Alliance's VC Interface

The Virtual Socket Interface (VSI) Alliance has proposed a standard for a VC interface that connects VCs to "wrapped" peripherals or on-chip buses (OCB) (contains logic to translate between the VC interface and the bus interface logic). The VC interface does not deal with processor buses or buses that are entirely internal to VCs; these are the responsibility of the VC developer.

Figure 5.6 illustrates the VC interface providing connectivity between the VC and the OCBs. The VC interface is the set of wires between the VC and the bus interface logic, which connects it to the bus. The darker boxes on either end of the VC interface are the logic necessary to create the VC interface on both pieces.

Figure 5.7 shows the VC interface in more detail. The low-speed, peripheral VC interface is a simple two-wire interface. The system bus interface or basic VC interface requires more complex control, and to use all the features of a complex system bus requires the full extensions of the interface. It needs a "wrapper" logic to connect the VC to the VC interface and a "wrapped" bus to contain logic to translate between the VC interface and the bus interface logic.

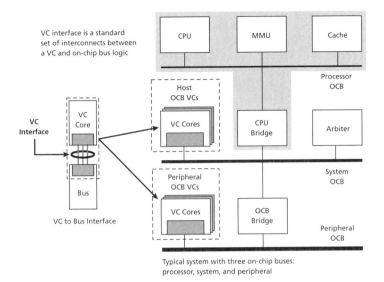

Figure 5.6. VC Interface

Much of this overhead logic will either be necessary because of the different address or data widths between the bus and the VC, or will disappear when the two components are synthesized together. The wrappers can be complex before the wrapper is synthesized. If the VC and bus are compatible, the wrappers should largely disappear. If the VC and bus are not highly compatible, we need a wrapper, which will have overhead (performance and gates), to make them work together. This overhead is offset by the ease of connecting any VC to any bus. The *VC Interface Specification* has options to accommodate most buses and VCs.

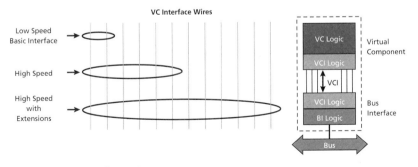

Figure 5.7. VC Interface Wires

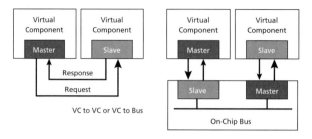

Figure 5.8. VC Connections

To keep the interface simple, the VC interface is a set of uni-directional point-to-point connections across the interface. There are two sides to the interface, a master and a slave. The master side makes requests and the slave side responds. An initiator would have a master VC interface, and a target VC would have a slave VC interface. The bus then must have both types of VC interfaces to provide each component the opposite interface to connect to. If a block is both an initiator and a target, it requires both types of VC interfaces and connects to two VC interface slots on the bus.

The VC interface is a simple request-response system. Each transaction is a pair of data packets. The write request contains words or "cells" of data, where a cell is the size of the data interface across the VC interface. A read request has no data, but the response packet contains the requested data. One or more cells of data can be transferred in each packet. Two pairs of symmetric request-grant signals control the two requests. The master side issues a request for the initial packet, and the slave side then issues a grant. The response packet is issued with a response request from the slave, and the master side then issues a response grant. The response packets contain an end of packet signal on the transfer of the last cell in the packet. For more details review the *VSI OCB Transactions Specification*.

Transaction Languages

Transaction languages describe the transactions from VC to VC across a bus. A transaction language can be written at several levels. These are syntactic ways of communicating transactions, generally similar to generic I/O commands in a high-level language. For example, the low-level VSI OCB transaction language[1] includes the following types of commands:

```
No operation to specified address. Return false if
error, true if no error.
bool vciNop (unsigned int address)
```

---

1. VSI Alliance's OCB VC Interface Specification, OCB Development Working Group.

```
Store 8, 16 or 32 bits. Return false if error, true if
not.
bool vciStore ([unsigned char *r_opcode,] unsigned char
p_len, unsigned int address, unsigned char mask, char |
short | int data[, bool lock])


Load 8, 16 or 32 bits. p_len based on r_data. e_data
contains returned data. Return error as above.
bool vciLoad ([unsigned char *r_opcode,] unsigned char
p_len, unsigned int address, unsigned char mask, char |
short | int *r_data[, char | short | int e_data] [, bool
lock])
```

These are relatively complex reads and writes. They include all of the parameters necessary to transfer data across a VC interface. A file syntax also exists to use with a simulation interface of the VC interface so that VCs can be tested either standalone or within the system, using the same data.

At this level, the transfer is not limited to the size of the cell or packet as defined in the VC interface. The specific VC interface information is defined in the channel statements, and kept in a packed data structure for the subsequent read and write commands. The following opens the channel:

```
int *vciOpen (int
*vci_open_data,"r"|"w"|"rl"|"rlw"|"rw",unsigned, char
contig, unsigned int p_len, unsigned int b_address, int
e_address [,unsigned char mask[, bool lock]]).
```

where vci_open_data is the control block that holds the data, followed by the type of transactions on the channel (r is read only, w is write only, rl is readlock, rlw is readlock write, and rw is read/write). b_address is the base address, and e_address is the end or upper address for the channel. The rest of the parameters are the same as used in the vciload and vcistore operations. This command returns zero or -error for unable to open. Possible errors include "wrong p_len" or "mask unsupported". In this case, the mask applies to all the data of cell size increments. vci_open_data should be 2 * address size + cell size + 10 bytes with a maximum of 64 bytes.

The following command writes a full transaction:

```
Int vciTWrite (int *vci_open_data, unsigned int address
unsigned int t_len, char *data[,unsigned char *mask[,
bool lock]])
```

where t_len is the amount of data to be transferred; mask is optional and anded with the open mask. vciTWrite stops on the first error. It returns the amount of data transferred.

The following command reads a full transaction:

```
int vciTRead (int *vci_open_data, unsigned int address,
int t_len, [char *r_data[, char *e_data[, unsigned char
*mask[, bool lock]]]])
```

where t_len is the amount of data to be transferred; mask is optional and anded with the open mask. If e_data is provided, the count is to the first bad cell of data. vciTRead stops on the first error. It returns the amount of data transferred.

The following command closes the channel:

```
int vciClose(int *vci_open_data)
```

vciClose returns the last error. An error will occur if the channel is not opened for the type of operation being done.

When invoked, these read and write commands can call the simpler packet load multiple times to complete the required transactions.

Designing Communications Networks

When designing communications networks, the buses must be defined, organized, and created in a way that corresponds to the communication requirements between blocks in a specific design. This section provides methods on how to determine the required bandwidths and latencies between blocks, what types of ports to create between buses and blocks, and what type of arbitration is appropriate for each bus, when using a block-based design (BBD) methodology.

Mapping High-Level Design Transactions to Bus Architectures

In the process of refining a design from the systems model, the highest point in which the communication between blocks in a design can be analyzed is at the cycle-approximate behavioral level. Cycle-approximate behavior provides capacity information on the function being described, such as the number of clock cycles required per operation. At this level, the communication between blocks is either direct signals or packets. The process below describes how to translate statistics acquired from simulating the design at this level into a definition of the necessary bus communication structure between the blocks.

The system design process described earlier started at an algorithmic, functional level. It successively refined the design's communications and functional partitioning to a level where adequate statistics can be extracted and used in defining the bus architectures for implementation.

Creating the Initial Model

Many system-level tools are capable of obtaining point-to-point bandwidth and latency information from the system design using high-level testbenches. A significant methodology transition is occurring from the current techniques to function-architecture co-design; but if these modeling methods are not available, an alternative modeling and statistics extraction technique can be used, which is described here.

First, we start with a block functional model of the chip, which contains functional models of the blocks and an abstract model of the interconnections between them, as well as testbenches consisting of sets of functional tests exercising a block or combination of blocks. Typically, this abstract interconnect model is a software mechanism for transferring data from a testbench and blocks to other blocks or the testbench. Ideally, it takes the form of a communication manager (and possibly scheduler) to which all blocks are connected. This scheduler is usually at the top level of the simulation module. The pseudo code for such a scheduler might look something like this:

```
While queue is not empty Do;
Get next transaction from queue;
Get target block from transaction;
Call Target Block(transaction);
End;
```

Where each block does the following:

```
Target Block(transaction);
Do block's function;
Add new transactions to the queue;
End
```

At this level there is no defined timing or bus size. All communication is done in transactions or as packet transfers. The packets can be of any size. The transactions can include any type of signals, since all communication between blocks goes through the scheduler. Alternately, the more direct, non-bus oriented signals can be set and read in a more asynchronous nature as inferred by the pseudo code below, which was modified from the block's code above:

```
Target Block(transaction);
Get direct, non-bus signal values from top level;
Do block's function;
Add new transactions to the queue;
Apply new direct, non-bus signal values to top level;
End
```

For simplicity, subsequent examples do not include these signals, but similar adjustments can be made in order to include non-bus type signals.

The testbenches should include sufficient patterns to execute the functionality of the entire chip. Target performance levels are assigned to each of the sets of patterns at a very coarse level. For example, if frame data for an MPEG decoder existed in one pattern set, the designer should be able to define how long the target hardware takes to process the frames in that set. In this case, the output rate should be equal to or greater than 30 frames per second, therefore the processing rate must exceed that number. These performance targets are used in the subsequent stages of this process to define the required bus bandwidths.

The selected blocks for the chip should have some cycle-approximate specifications. These either already exist within the block functional models, or they need to be incorporated into the model in the next step.

Modifying the Interconnect Model

Some designs, such as hubs and switches, are sensitive to data latency. Most network devices, especially asynchronous transfer mode (ATM), have specific latency requirements for transferring information. If a design has no specific latency requirement, it is not necessary to add cycle count approximates to the model.

To adjust the interconnect model or scheduler, which transfers the data from one block to another, we first need to add the amount of data that is being transferred from one block to another and the number of transactions that are conducted. This data is accumulated in two tables for each pattern set. For example, in a chip with three blocks and a testbench, each table would be a 4x4 from-to matrix with the sum of all data transferred (in bytes) in the first table, and the count of all transactions in the second table (see Table 5.2 and Table 5.3). The diagonal in both tables should be all 0s. A more practical model should also consider the buses going into and out of the chip, so the testbench would probably have more than one entry on each axis.

Table 5.2. Data Transfer for Pattern Set X

From/To	Block 1	Block 2	Block 3	Memory	PCI	PIO
Block 1	0	10,000	100	10,000	100	100
Block 2	8,000	0	100	2,000	10,000	100
Block 3	200	100	0	100	200	100
Memory	6,000	6,000	100	0	100	100
PCI	6,100	4,100	0	200	0	0
PIO	0	0	0	0	0	0

Table 5.3. Transactions for Pattern Set X

From/To	Block 1	Block 2	Block 3	Memory	PCI	PIO
Block 1	0	200	25	200	4	25
Block 2	160	0	100	40	200	25
Block 3	25	100	0	25	25	25
Memory	120	180	25	0	6	25
PCI	122	82	0	50	0	0
PIO	0	0	0	0	0	0

These tables were created using the following pseudo code:

```
While queue is not empty Do;
Get next transaction from queue;
Get sender block from transactions;
Get target block from transaction;
Get Transaction byte count;
Transactions   Matrix  (sender,target)   =Transactions
Matrix(sender,target) + 1;
Data Transfer   Matrix  (sender,target)   =Data Transfer
Matrix(sender,target) +
Transaction byte count;
Call Target Block(transaction);
End;
```

In the second step, the blocks have their estimated clock cycles per operation added to the existing block functional models. The block models need to be modified to reflect the cycle-approximate operation as defined by their specifications if they do not already reflect the specification's operation. This would typically be done before layout of the block, but after completion of behavioral verification of the block's function. The cycle time of the clock should have already been defined, in order to translate raw performance into cycle counts. After the approximate cycle times have been added to the block's functional models, they should be integrated back into the chip model. This model will have cycle-approximate blocks with no delay in the interconnect. A table similar to the ones above is then set up, but this time it should contain the number of cycles each transfer should take, from the time the data is available to the time the data arrives at the next block or testbench. The interconnect model should then be modified to use this table. The pseudo code for these modifications is:

```
While queue is not empty Do;
Get next transaction from queue;
Get time from transaction;
Get target block from transaction;
Call Target Block(transaction, time);
End;
```

Where each block does the following:

```
Target Block(transaction,time);
Do block's function;
Set Transactions' times to time + delay + Latency(this
block, target);
Sort new transactions to the queue;
End
```

Block to block signals are added as separate transactions in the timing queue, in addition to the bus transactions, since these signals also have some delay (typically at least one clock cycle).

The testbench can then be modified to include the chip latency requirements. At this point, the designer needs to add estimated interconnect cycle count delays, based on the flow of data in the design. The design is then simulated to check whether it meets the cycle requirements of the design. Modifications are then made to the table, and the verification process is repeated until the cycle requirements of the chip are met. The designer should use large interconnect delays to start and reduce them until the specifications are met, which creates a table with the maximum cycle counts available for each type of bus transfer. These tighter latency requirements translate into more gate-intensive, bus-interconnect schemes. Table 5.4 shows an example of a latency matrix.

Table 5.4. Latency Matrix for all Pattern Sets

From/To	Block 1	Block 2	Block 3	Memory	PCI	PIO
Block 1	na	50	1,000	50	100	1,000
Block 2	50	na	1,000	300	100	1,000
Block 3	1,000	1,000	na	500	500	1,000
Memory	50	50	500	na	100	1,000
PCI	100	100	na	50	na	na
PIO	na	na	na	na	na	na

Cells that contain "na" in Table 5.4 indicate that no data is transferred, and therefore are not applicable to the latency matrix.

Alternatively, this latency and bandwidth information can be obtained more directly through the function-architecture co-design methodology and tools. These tools include the monitoring and scheduling alternatives that are described in the pseudo-code examples above.

Now having created the initial matrices, the subsequent procedures are applicable in either case.

Transforming Matrices

The data matrix must now be transformed to reflect the natural clustering of the data. This clustering transformation is done by trying to move the largest counts closest to the center diagonal. There are a number of ways clustering can be done; the process described below is one such way.

We now need a method for evaluating the "goodness" of the clustering. A goodness measure is the magnitude of the sum of the products of each data transfer count times the square of the distance that cell is from the diagonal. In other words, the measure is the sum of the products of all cells in the data transfer matrix and a corresponding distance measure matrix. For the 6x6 matrices described above, a distance matrix might look like Table 5.5.

Table 5.5. Distance Measure Matrix

From/To	Site 1	Site 2	Site 3	Site 4	Site 5	Site 6
Site 1	0	1	4	9	16	25
Site 2	1	0	1	4	9	16
Site 3	4	1	0	1	4	9
Site 4	9	4	1	0	1	4
Site 5	16	9	4	1	0	1
Site 6	25	16	9	4	1	0

Other measures could be used, but the square of the distance converges quickly while allowing some mobility of elements within the system, which higher-order measures would restrict.

Now sort the sites as elements:

```
Get Current cluster measure of matrix;
Do for Current site = site 1 to n-1 in the matrix;
Do for Next site = Current site +1 to n in the matrix;
    Swap Next site with Current site;
    Get Next cluster measure of matrix;
    If Next cluster measure > Current cluster measure
Then
            Swap Next site with Current site back to
original location.
    Else
            Current cluster measure = Next cluster
measure;
End
End;
```

This is similar to a quadratic placement algorithm, where interconnect is expressed as bandwidth instead of connections. Other methods that provide similar results can be used. With the method used here, the cluster measure of the original matrix is 428,200, and pivoting produces the matrix shown in Table 5.6 with a cluster measure of 117,000.

Blocks 1 and 2, which have high data rate communication with the PCI and Memory, must be on a high-speed bus, while the block 3 and PIO can be on a low-speed bus. The PIO provides output only where all the others are bi-directional. Also, because there is no communication between the components on different buses, a bridge is necessary. We have defined the bus clusters, but not the size and type of bus. In this example, no information is created, so what

Table 5.6. Pivoted Data Transfer Matrix

From/To	PCI	Block 2	Block 1	Memory	Block 3	PIO
PCI	0	4,100	6,100	200	0	0
Block 2	10,000	0	8,000	2,000	100	100
Block 1	100	10,000	0	10,000	100	100
Memory	100	6,000	6,000	0	100	100
Block 3	200	100	200	100	0	100
PIO	0	0	0	0	0	0

is read is written, hence each column and row totals match (except for Block 3 and PIO). This is not usually the case.

Selecting Clusters

With predefined bus signals, the initial clustering is done for all the connections defined for those signals. This is pivoted to show the natural internal clusters, but the original bus connections are still considered as one cluster, unless more than one bus type is defined for the signals. In that case, the processor's system and peripheral buses are defined. The cluster is then broken into a system bus and peripheral bus or buses, based on the clustering information. For example, if the bus matrix defined in Table 5.6 were for a predefined set of buses, the initial clustering would be for the whole matrix. But if more than one bus was defined, the blocks that need to be on a high-speed bus would form one bus and the rest would form another. This partition is then passed on to the next step.

In the rest of the cases, no predefined bus connections exist. These need to be divided up based on the cluster information. Typically, the pivoted matrix has groups of adjacent blocks with relatively high levels of communication between them, compared to other adjacent blocks.

For example, in Table 5.7, A, B, and C form one independent bus cluster, because there is high communication among them. There is no communication between A, B, and C and blocks D through H. Blocks D, E, and F form another cluster, because they have high communication. The DE and EF pairs could form two separate buses—a point-to-point for DE and a bus for EF. GH is a third cluster. There are lower bandwidth connections between the EF pair and the GH pair. Again, depending on the amount of intercommunication, the four blocks, EFGH, could be on one bus, or they could be on two separate EF and GH buses with a bi-directional bridge between them for the lower level of communication.

Table 5.7. Abstract Pivoted Matrix

From/To	A	B	C	D	E	F	G	H
A	0	##*	##					
B	##	0	##					
C	##	##	0					
D				0	##			
E				##	0	##		
F					##	0	6	5
G						6	0	##
H						5	##	0

*## denotes a large number

Table 5.8. Abstract Pivoted Matrix with One Cut

From/To	A	B	C	D	E	F	G	H
A	0	##	##					
B	##	0	##					
C	##	##	0					
D				0	##			
E				##	0	##		
F					##	0	6	5
G						6	0	##
H						5	##	0

Cluster identification requires some guidelines on how to choose from a number of different options. Let's start with identifying the cut points between the blocks to determine the possible clusters. A cut point is where less communication takes place across the cut than between blocks on either side of the cut. Using the Abstract Pivoted Matrix in Table 5.7, a cut between C and D would produce the diagram in Table 5.8.

The communication between the two groups, ABC and DEFGH, is defined by the sum of all the cells in the lower left quadrant plus all the cells in the upper right quadrant. If this sum is 0 (which it is this case), the two groups have no communication between them and will form completely separate buses. So first cut the pivoted matrix where the resulting communication across the cut is 0.

Next, within each of the identified quadrants find the non-trivial cuts. A trivial cut is one block versus the rest of the quadrant. The cuts should be significant, meaning the communication between the resulting groups should be much less than within each group.

In the Abstract Pivoted Matrix in Table 5.7, the first quadrant has no cuts, and the second quadrant has one, as shown in Table 5.9. Here, the communication between the lower two quadrants is 22, where the communication within each of the quadrants is a very large number (##). This could indicate two buses with a bridge between them.

If this technique is employed on the original example (Table 5.6), the clusters in Table 5.10 are created. This example shows two buses with a bridge between them. One has a lot of data transferred on it, while the other has very little. Another cut between Block 3 and PIO would have resulted in an even lower communication between the clusters, but this is a trivial cut because it leaves only one block in a cluster, and was therefore not used.

This technique does require system knowledge. The timing of the data and the implementation details, such as existing bus interfaces on blocks, the addi-

Table 5.9. Abstract Pivoted Matrix with Multiple Cuts

From/To	A	B	C	D	E	F	G	H
A	0	##	##					
B	##	0	##					
C	##	##	0					
D				0	##			
E				##	0	##		
F					##	0	6	5
G						6	0	##
H						5	##	0

tional requirements of a processor, and the number of masters on the bus, are outside the scope of this procedure, but should be taken into consideration. By deviating from the cluster structure obtained by the methods described here, a bus structure that has either better performance or lower gate count could be created. In that case, when these factors are determined, the developer might want to return to this method to modify the clustering results.

Selecting Bus Types and Hierarchy

The next step is to define the attributes of each of the buses identified in the clustering process described previously. To select the appropriate bus, each cluster is analyzed for existing bus interfaces. If none or few exist, the bus is selected by matching the attributes of buses available in a user library. The outputs of this process are a defined set of buses and a bus hierarchy, which are used in the next step.

Buses can be categorized according to latency and bandwidth utilization, which is a function of architecture. Pure bandwidth is a function of the number of wires in the bus times the clock frequency the data is being transferred at. Table 5.11 lists bus attributes from lowest bandwidth utilization and longest

Table 5.10. Pivoted Data Matrix with Clustering

From/To	PCI	Block 2	Block 1	Memory	Block 3	PIO
PCI	0	4,100	6,100	200	0	0
Block 2	10,000	0	8,000	2,000	100	100
Block 1	100	10,000	0	10,000	100	100
Memory	100	6,000	6,000	0	100	100
Block 3	200	100	200	100	0	100
PIO	0	0	0	0	0	0

Table 5.11. Bus Taxonomy Table

	Bandwidth Utilization		Latency	
Bus Type	Min	Max	Data	Transfer
1) Serial, asynchronous, clock-regenerated bus	5%	25%	50	200
2) Multiple line, asynchronous, clock-regenerated bus	5%	25%	20	100
3) Multiple line, combined data-address synchronous bus	10%	25%	5	25
4) Separate data and address synchronous bus	25%	50%	2.5	10
5) Single-level pipelined data and address bi-directional bus	25%	75%	2	5
6) Multiple-level pipelined bus with sophisticated arbitration	50%	75%	1.5	2.5
7) Cross-bar switch	75%	100%	1	2
8) Point-to-point uni-directional wire	100%	100%	0.5	1

latency to the highest bandwidth utilization and shortest latency. Typically, the cost in logic and wires is smallest with the first, and largest with the last.

Bus type is defined by a range of latency (cycles) and bus bandwidth (utilization percentage). Each bus can have a different clock cycle time and size. The utilization percentage is the effective throughput divided by the product of the cycle time times the size of the bus; 100 percent means every cycle is fully utilized. The Latency Data column is the number of cycles needed for a bus word of data to be transferred. The Transfer column is the average number of cycles to begin a bus transaction.

A library of buses gets created after a number of projects. Each bus entry should contain information on the bus type and attributes from the VSI Alliance's *OCB Attributes Specification*. Some examples of bus types are PCI, which is a type 4 bus, and AMBA's system bus, which is a type 5. The board-level bus for the Pentium II is a type 6 when used in a multiple processor configuration.

Bus Clustering Information
Next, the bus latency, bandwidth, and clustering information needs to be translated into a form that is useful for determining the type and size of the

Table 5.12. Reduced Bus Matrix

From/To	Bus 1	Bus 2
Bus 1	62,600	600
Bus 2	600	100

buses. If we look at the information in Table 5.10, the first four entries are clustered in one block, and the last two are clustered into a second block. The bus bandwidth is first determined by summing up all the transactions that occur within the identified clusters in the matrix. In Table 5.10, this is 62,600 within the large cluster, 100 within the small cluster, and 1,200 between the clusters, as shown in Table 5.12, which is created by summing all the entries in the four quadrants.

For example, if the time this pattern set is expected to take is 1 millisecond, the fast cluster must transfer 63,800 bytes of data in 1 millisecond—1,200 bytes to the bridge and 62,600 bytes internal to the bus. This translates to a 510 megahertz bandwidth. If the clock cycle is 20 nanoseconds, and the bus utilization is 25 percent, the number of bits rounded to the nearest power of 2 is 64. Or 64 * 25%/20ns = 800 mhz > 510mhz. If we use a type 4 or 5 bus, we need at least 64 bits. With a 20-nanosecond cycle time, we need only 8 bits for the slower cluster.

Latency information is partially a function of the utilization, because increased utilization of a bus causes increased latency. This complexity has not been included in this example, since it is partially accounted for in the utilization numbers. But assuming we use the minimum bus utilization numbers for the bandwidth calculation, the latency should be toward the minimum as well. To create a margin, we should select the worst case latency requirement (smallest) from the cluster. The latency matrix in Table 5.4 provides the latency of the entire transaction, but the Bus Taxonomy Table has the bus latency data and transfer as separate numbers. For example, for a type 4 bus, the transfer latency is 10. The data latency is the number of cycles required for the data alone. We have to calculate what the transfer latency would be by subtracting the data transfer time from the numbers in the latency matrix. The data transfer time is the data latency cycles for this bus type divided by the number of words in the bus times the average transaction size. The average transaction size is the number of bytes of data from Table 5.2 divided by the number of transactions in Table 5.3. To compare the latency from the table, we have to make a latency matrix as shown in Table 5.13, which is based on the latency matrix from simulation (Table 5.4) minus the transaction's data latency.

Table 5.13. Pivoted Resulting Latency Matrix

From/To	Block 1	Block 2	Block 3	Memory	PCI	PIO
Block 1	na	31	988	31	91	988
Block 2	31	na	997	281	81	988
Block 3	976	997	na	488	476	988
Memory	31	38	488	na	94	988
PCI	81	81	na	49	na	na
PIO	na	na	na	na	na	na

Each element in this matrix is calculated as follows:

```
Resulting Latency(x,y) = Latency (x,y) - Bus Latency
data(type) *
Data Transfer(x,y) / [Transaction(x,y) * bus size]
```

The smallest number in the system bus cluster is 25. This should be larger than the transfer latency for the type of bus we need because of bandwidth. In the Latency Transfer column of the Bus Taxonomy Table that number is 10, bus type 4. We can therefore choose a bus type 4 or better for the fast cluster.

Selecting Buses
Selecting buses is typically done using the following steps:

1. Eliminate buses that do not meet the cluster's bandwidth and latency requirements.
2. If the bus is already defined, use that bus; otherwise go to step 3.
3. If a processor is present, use a system bus that it already connects to; otherwise go to step 4.
4. Select a bus most blocks already connect to.
5. Use a bus that can handle the endian method of most of the blocks connected to it.
6. Use multiple buses if the loading on the bus is excessive.
7. Separate out the lower bandwidth devices onto a peripheral bus or buses.
8. Use a peripheral bus that has an existing bridge to the selected system bus.

Each of these conditions can be tested by inspecting the parameters in the bus library and the interfaces of the blocks in the design. If there is more than one choice after this selection process, choose the one that best meets the VSI Alliance's OCB Attributes list (this will be the one with the most tool and model support, etc.).

After the buses and their loads are identified, the bridges need to be identified. If two buses are connected in the reduced bus matrix in Table 5.12 (their from/to cells have non-zero values), a bridge must be created between them. Using the pivoted data matrix and the reduced bus matrix, we can create the following bus model:

System bus (type 4 or 5) of 64 bits connected to:
    Block 1 (R/W)
    Block 2 (R/W)
    Memory (R/W)
    PCI (R/W)
A Bridge (R/W) to:
    Peripheral bus (type 3 or better) of 8 bits connected to:
    Block 3 (R/W)
    PIO (Write only)

The PIO is write only, because no data comes from it. The bridge is read/write, because both diagonals between bus 1 and 2 are non-zero. This model is used in the next task.

Creating the Bus Design

In this step, the selected buses are expanded into a set of interface specifications for each of the blocks, a set of new blocks, such as bridges, arbiters, etc., and a set of remaining glue logic. The block collars and new blocks are implemented according to the specifications, and the glue logic is transferred as mini-blocks to chip assembly.

Defining the Bus Structure

In defining the bus structure, we can first eliminate all buses with a single load and a bridge by putting the load on the other side of the bridge. It is both slower and more costly in gates to translate between the protocol of the system bus and the peripheral bus for only one load. The bridge logic cannot be entirely eliminated, but the tristate interface can. The peripheral bus reduces to a point-to-point communication, and its 8 bits can be turned into 16 without much penalty.

Next, we need to assign bus masters and slaves to the various loads. We can start with the bridge. The slower peripheral side has a master, the faster system side a slave. All devices on peripheral buses are slave devices. On the system bus, the master and slave are defined by which devices need to control the bus. If a processor is connected to the bus, its interface is a master. Otherwise, if there are no obvious masters, the external interface, such as the PCI, is the master. The memory interface is almost always a slave interface. To determine which block requires a master interface, refer to the bus's interconnect requirements.

If a processor or other block is connected to a bus that has a memory interface, and the block specifically requires it, include one or more direct memory access (DMA) devices on the bus to act as bus masters. If there are two or more bus masters, add an arbiter.

Creating the Detailed Bus Design

With the structure defined, the detailed bus interface logic must now be created. If the interfaces already exist on the blocks, they should be in a soft, firm, or parameterized form, so they can be tailored to the bus. If this is the case, use the existing bus interface logic; otherwise use the models provided with the bus. If the blocks have a different bus interface, eliminate it if possible. The bus interface logic is then connected to the resulting interface of the block. This bus interface logic must be modified so that it interfaces with the bus, as follows:

1. Assign address spaces for each of the interfaces.
   The address space is usually designed to match the upper bits of the transaction address to determine whether this block is being addressed. Make sure that each block has sufficient address space for the internal storage or operational codes used in the block.
2. Eliminate write or read buffers if only one function is used.
   Most existing bus interfaces are designed for both reads and writes. If only one direction is needed, logic is significantly reduced. For example, if the bus takes more than one clock cycle, read and write data is usually separately buffered. If only one direction is needed, half of the register bits can be eliminated.
3. Expand or contract the design to meet the defined bus size.
   Most existing bus interfaces are designed for the standard 32- or 64-bit bus, but other alternatives are often available. This requires eliminating or adding the extra registers and signal lines to the logic. For buses that interleave the address and data onto the same bus signals, a mismatch in data and address size eliminates only the upper order address decode or data register logic, not the data signals.
4. Modify the bridges' size mappings between their buses.
   This is the same as step 3, but for both sides of the bridge.
5. Add buffers as necessary to the bridges.
   Bridges require at least one register for each direction be equal to the larger of the buses on either side for a read/write interface. In addition to the one buffer for data in each direction, bursts of data might be transferred more efficiently if the data is accepted by the bridge before being transferred to the next bus. This could require a first-in first-out (FIFO) memory in each direction where a burst is stored and forwarded on to the next bus, as shown in Figure 5.9.

16 bits   8 bits

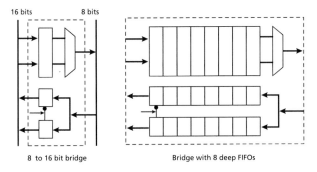

8 to 16 bit bridge   Bridge with 8 deep FIFOs

Figure 5.9. Bridges with FIFOs

6. Define the priority of the bus masters and type of arbitration.
   If more than one master on a bus exists, arbitration must occur between the masters. If the masters handle the same amount of data, with similar numbers of transactions and required latency, they should have equal polling priority. However, if there is a clear ranking of importance among the masters, an equivalent order for the amount of data, transactions, and lowest latency, the arbitration should be serial with the most critical master first.
7. Create and connect the arbiter based on the definitions in step 6.
   Arbitration schemes can be distributed or centralized, depending on the bus. Try to distribute the arbitration logic as much as possible, since it needs to be distributed into the blocks with the glue logic.
8. Map the bus to the interface logic as required by the device's endian method.
   While most buses are little endian, some devices are big endian. When different endian types are used, you must decide how to swap the bytes of data from the bus. Unfortunately, this is context-dependent in the most general case. If all transactions to and from the bus are of the same type of data, a fixed byte swapping can be employed, otherwise the bus masters must do the swapping.
9. Tailor any DMA devices to the bus.
   DMA devices, which are essentially controllers that transfer data from one block to another, must be modified to the size of the address bus.
10. Add any testability ports and interfaces, as necessary.
    The test features might require additional signals to differentiate the test from the normal operation mode.
11. Add any initialization parameters, as necessary.
    Some buses, such as PCI, have configuration registers, which can be hard-coded for those configurations that do not change.

12. Add optional bus capabilities as required by the devices on the bus. Some buses have advanced capabilities, such as threads, split transactions, and error retry, which might not need to be implemented if the devices connected to the bus do not require them. Some of the additional capabilities, such as DMA devices, non-contiguous burst transfers, and error recovery control, might require more signals than defined in the standard bus. These signals should be added to the bus, if necessary.

Port Splitting and Merging

In the example in the previous section, we assumed each block required only one interface or port to a single bus. This is not always the case. Under certain conditions, it is desirable to convert a single port into two ports, or a block that was designed with two ports into one that has only a single port. This is called port splitting and merging.

Port Splitting

Port splitting is done when there is a high point-to-point bandwidth or tight latency requirement between two blocks and one of the blocks only communicates with the other. Using the previous clustering example, as shown again in Table 5.14, if the communication between or within clusters is not between all blocks, some further optimization can be done. Optimization is necessary if the latency matrix has very different communication requirements between certain blocks. For example, the matrix shows that the GH cluster does not communicate with DE. Furthermore, DE and EF communicate but D and F do not. If the latency requirements for DE are very tight, it makes sense to split out the DE communication from the rest of the bus. The resulting matrix would look like the one in Table 5.15.

In this case, we split out E and E' into what appears as two separate blocks, because separate interfaces will be created on E for the two buses. If a block

Table 5.14. Abstract Pivoted Matrix with Multiple Cuts

From/To	A	B	C	D	E	F	G	H
A	0	##	##					
B	##	0	##					
C	##	##	0					
D				0	##			
E				##	0	##		
F					##	0	6	5
G						6	0	##
H						5	##	0

Table 5.15. Matrix with Split Ports

From/To	A	B	C	D	E	E'	F	G	H
A	0	##	##						
B	##	0	##						
C	##	##	0						
D				0	##				
E				##	0				
E'						0	##		
F						##	0	6	5
G							6	0	##
H							5	##	0

starts out with two or more bus interfaces, this technique can be used to effectively use the separate interfaces. Now the DE interface can be reduced to a point-to-point connection to satisfy the tight latency requirements. E' and F then form a bus with a bridge to the bus containing G and H.

Port Merging

Port merging is done when a block has two separate ports, and it is necessary to include both ports to create a proper set of data and latency matrices. The ports would then be sorted as if they were separate blocks. If the resulting clustering showed the two ports on two separate buses, they could be built that way, or if both buses are underutilized, they can be merged together. A process similar to that of merging of peripheral buses should be followed, but one target frequency of the merged bus must result, even if there were originally different clock frequencies for the two buses.

If the original ports consisted of an initiator and a target port, there might be little reduction in the resulting control logic. Figure 5.10 shows an example of

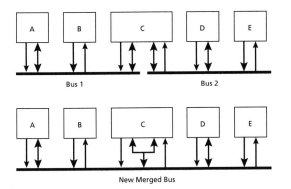

Figure 5.10. Port Merging

this kind of merging, where block C has an initiator on Bus 1 and a target on Bus 2. After merging the buses, the data port can merge but the address logic remains largely separate. This port merging can result in internalizing some arbitration between the original data ports, as is shown by the two data port interfaces remaining on block C.

Mapping Arbitration Techniques

Most arbitration techniques are generally mixes of round robin polling and serial priority schemes. Polling gives each of the initiators equal priority. It shifts from one to the next on each clock cycle, stopping when one of them wants the bus. Priority arbitration gives the bus to the initiator with the highest priority; the lowest priority initiator gets the bus only when all the other initiators do not require it.

To determine which arbitration structure makes the most sense, establish the minimum transaction latencies for each of the initiator blocks from and to all the other non-initiator blocks. Do the same for the transfer latencies. Sort this list by size of the transaction latency. This should produce a list of the initiator blocks in increasing latency, as shown in Table 5.16.

If the transaction latencies are all approximately the same size and close to the latency of the selected bus, choose a round-robin polling structure.

If the latencies increase successively as you go down the list by at least the transfer latency of the previous element in the list, or the minimum latency is larger than the sum of the bus's transaction latency plus all the transfer latencies of the initiators, choose an ordered priority arbitration structure. An ordered priority arbitration structure is less costly in gates and should be chosen when any arbitration structure would work. In addition to the latency requirements, the total required bandwidth on the bus must be sufficiently lower than the available bandwidth to allow lower priority devices access to the bus when using the ordered priority structure. In general, the deeper the priority chain, the larger the excess bandwidth must be.

Table 5.16. Latency Tables for Arbitration

Round Robin			Ordered Priority		
Initiator Block Name	Transaction Latency	Transfer Latency	Initiator Block Name	Transaction Latency	Transfer Latency
Block D	12	6	Block D	12	6
Block A	14	8	Block A	20	8
Block C	15	5	Block C	28	5

Using a Platform-Based Methodology

This section describes the communication differences between the block-based design methodology described above and a platform-based design (PBD). To develop derivatives quickly, you must start with a predefined core for the derivative that includes the processor, system bus, and the peripherals necessary for all products in the given market segment. These cores are called hardware kernels, and at least one is contained in each platform-based derivative design. The communication structure of a platform design is similar to the structure of the general design described above, which has predefined buses, but the bus structure of the design is separated by VC interfaces on the hardware kernel.

Mapping to a Platform

Using the *VC Interface Specification,* which includes a transactions language as well as the VC interface definition, you can define a specific bus structure within a hardware kernel and interface to it using VC interfaces. This enables the designer to specify the communication requirements as in the methods described earlier, but also deal with the portions of the design applied to the hardware kernel as predefined bus interconnections.

The major difference between the general methodology described earlier and PBD is the way the bus is treated after the transaction analysis is completed. In the general mapping process, the bus components must be developed and inserted in the design as separate mini-blocks. In PBD, the bus starts out as a predefined part of the hardware kernel, as shown in Figure 5.11.

The hardware kernel has VC interface connections to the other blocks in the design. To get from the initial form to the translated form, you must execute a modified form of the earlier methodology, but unlike that methodology, the blocks in PBD can be either software or hardware. The software blocks are allocated within the processor block or blocks contained within the hardware kernel. Once this assignment is completed, a cycle-approximate behavioral model

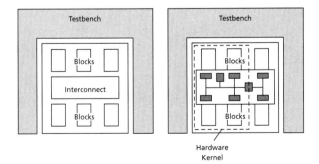

Figure 5.11. Conversion from Interconnect to Bus Structure

of the blocks, including the processor block with the allocated software blocks, is created. The communication in this model occurs between the blocks within the model in the same manner as in the general communication model.

With the more formal function-architecture co-design approach, mapping can be done using an algorithm similar to the one described here or through other iterative techniques.

Clustering can be done in the same way. Since the blocks within the hardware kernel are already connected to specific buses, they should either be modeled as one block or moved together in the clustering process. The end result should show which blocks in the derivative design belong on each bus. This is defined using the following procedure:

1. Define one block for each bus internal to the hardware kernel.
2. Include all the blocks on that bus within each created block.
3. Delete the internal block to block bandwidth/bus utilization from each bus.
4. Add each peripheral block to the matrices.
5. Pivot the matrix as defined in the general method.
6. Assign all the peripheral blocks to the hardware kernel bus blocks in order of the highest affinity first, up to the VC interface or bandwidth limits of each bus.
7. If there are more peripheral blocks than VC interfaces or available bandwidth on a hardware kernel bus block, create a bus with a bridge connected to one of the hardware kernel bus block's VC interfaces and reorder the peripheral blocks according to their clustering in the pivoted matrices.
8. Connect the peripheral blocks to their assigned bus or VC interface.

In this procedure either the peripheral blocks are all assigned to VC interface ports on the hardware kernel, or one or more additional external buses will be created. If the clustering suggests no additional buses need to be created, the assignment can be implemented as shown in Figure 5.12.

If additional buses need to be created, connect the appropriate blocks to it. The additional external bus is then connected by a bridge to one of the hardware kernel's VC interfaces, which, in turn, is connected to the hardware kernel's system bus.

During this process, add arbiters or bridge logic, as necessary, depending on which blocks are initiators and which are targets. In general, the initiator blocks should all be either connected directly to the hardware kernel's system bus via a VC interface, or a bi-directional bridge with initiator and target capability should connect the hardware kernel's system bus to an external system bus containing an arbiter. This type of bridge requires both a master and a slave interface to the hardware kernel's bridge. If this displaces an additional peripheral block, assign that block to the next closest bus in the sorted matrix.

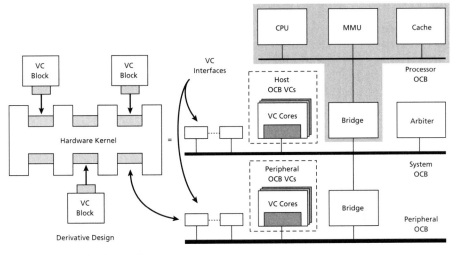

Figure 5.12. Derivative Design

Verifying the Bus Structure

To test each of the blocks the vectors need to be in transaction language form. Each block is individually tested first with its vectors. Then later, the testbench is used to communicate in the transaction language through the bus to the individual blocks. The vectors are distributed in the same fashion that would be seen in the system. This could be the initial system bus verification method, which later can be augmented with system-level transactions.

The transaction language is hierarchical. The highest level is timing-independent, while the lowest level is cycle-timing specific. This results in a new methodology for migrating the testbench, so that it can be applied to successively more accurate models, while keeping the same functional stimulus (this is further discussed in Chapter 7).

Bus Mapping Example

Assuming we have the following hardware kernel:

System bus (type 4 or 5) of 64 bits connected to:

Processor (R/W)
VC interface (R/W)
VC interface (R/W)
PCI (R/W)
A bridge (R/W) to peripheral bus (type 3 or better) of 8 bits connected to:
VC interface (R/W)
VC interface (R/W)
PIO (Write only)

Further assume we have blocks A through H in a design. Blocks A through E are software blocks, and F through H are hardware blocks. Since there is only one processor, assign blocks A through E to the processor block. Simulation yields the following data transfer matrix:

From/To	Processor	Block F	Block G	Block H	PCI	PIO
Processor	0	10,000	100	10,000	100	100
Block F	8,000	0	100	2,000	10,000	100
Block G	200	100	0	100	200	100
Block H	6,000	6,000	100	0	100	100
PCI	6,100	4,100	0	200	0	0
PIO	0	0	0	0	0	0

Collapsing the hardware kernels to one block per bus produces the following:

From/To	Block X	Block F	Block G	Block H	PIO
Block X	0	14,100	100	10,200	100
Block F	18,000	0	100	2,000	100
Block G	400	100	0	100	100
Block H	6,100	6,000	100	0	100
PIO	0	0	0	0	0

We created block X for the system bus, and put the processor and PCI block into block X. The total number of bytes transferred internally to the PCI and processor is 6,200 bytes. A type 4 or 5 bus's minimum utilization is 25 percent (see Table 5.11), so the reduction on the required bandwidth must be the actual data transferred (6,200 bytes) divided by the utilization (25 percent). In other words, if only one byte in four is used on the bus, the reduction in required utilization is four times the number of bytes no longer transferred on the bus. We do not need to do anything with the PIO, because there are no other blocks to merge with.

Now, we can pivot the lower matrix, which yields the following:

From/To	Block F	Block X	Block H	Block G	PIO
Block F	0	18,000	2,000	100	100
Block X	14,100	0	10,200	100	100
Block H	6,000	6,100	0	100	100
Block G	100	400	100	0	100
PIO	0	0	0	0	0

Since there are two VC interfaces for the bus block, X, F, and H connect to those VC interfaces. Similarly, block G connects to one of the VC interfaces on the peripheral bus. If there were only one VC interface on the system bus (which in actuality cannot happen because of the need for verification), a bridge block would need to be inserted with another system bus for blocks F and H. This structure looks like the following:

> System bus (type 4 or 5) of 64 bits connected to:
> > Processor (R/W)
> > PCI (R/W)
> A bridge (R/W) to another system bus (type 4 or 5) of xx bits con-

nected to:
> > Block F (R/W)
> > Block H (R/W)
> A bridge (R/W) from the original system bus to peripheral bus (type

3 or better) of 8 bits connected to:
> > Block G (R/W)
> > VC interface (R/W)
> > PIO (Write only)

There are only two VC interfaces, so testing consists of connecting the behavioral VC interface model to one of the interfaces, and connecting block F to the other. After testing it, the locations are swapped, so that block H can have the slot previously occupied by the behavioral VC interface model, and the behavioral model can be installed in block F's site. Note that because the address has to be compensated for in the VC interface model, the transaction vectors need to be relocated.

Communication Trade-offs

This section discusses some of the communication trade-offs that occur in regards to memory sharing, DMA and bridge architectures, bus hierarchy, and mixing endian types.

Memory Sharing

At the algorithmic design level, there is no separation between software and hardware design. As the functionality is broken out into separate sections or blocks in the design, transferring information between the blocks is typically done through memory. For example, one task gets some information from memory, transforms it in some fashion, and puts it back into memory. The next task, or the task after that, gets these results and further processes them. When

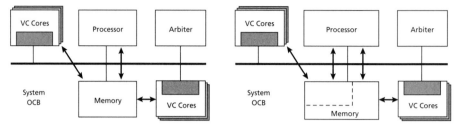

Figure 5.13. Memory Sharing

converted to a behavioral-level design, this leads to a shared memory structure. The simplest structure is one memory with all devices accessing it as can be seen in the left diagram in Figure 5.13. By separating the memory structures into distinct areas corresponding to the communication between different blocks, the memory can appear as a holding area for data that is being transferred between blocks, as illustrated in the right diagram in Figure 5.13.

This can be further refined by separating the memory where bandwidth and clustering of the blocks warrants separate structures. These are separate blocks of memory, each being shared by two or more blocks, as a way to communicate blocks of information that are larger than a single burst of data. Clustering may result in a separate bus. The shared memory communication structure can be converted into a bridge with memory, or a FIFO to store this stream of data between separate blocks and buses in the design, as shown in Figure 5.14.

In the simplest case, communication might be between memory and two other devices. If one is serially writing the data and the other is serially reading the same data, the memory can be replaced by a simple FIFO, and no intervening bus is necessary. If this serial transmission is occurring between multiple blocks on separate buses, a FIFO bridge might be appropriate. If sev-

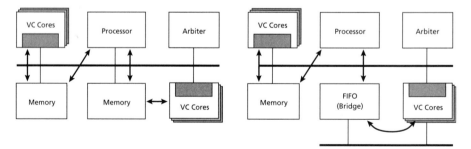

Figure 5.14. Separating Memory Blocks

eral are reading and writing, an equivalent number of FIFOs might be appropriate. If the blocks are randomly reading or writing data to the common memory and no amount of further segmentation can separate out these random interactions, the communication between the blocks should be through a shared memory.

Whenever possible, convert shared memory to any type of FIFO, because on-chip memory is limited relative to off-chip memory. Access to off-chip memory takes many more clock cycles than on-chip memory. If it is possible to convert, you must determine the size of the FIFO, which is discussed in the next section.

FIFO Design and Depth Calculation

Figure 5.15 shows the basic structure of memory-based and register-based FIFOs. The memory-based FIFO is a 1R/1W two-port memory, with a write and read counter. It also contains some comparison logic to detect an underflow (when the read counter equals the write counter) or overflow (when the write counter counts up to the read counter). The read and write counters start at zero and increment after each operation. They are only as large as the address bits of the memory, so they automatically wrap from the largest address back to zero.

The register FIFO captures the data in the farthest empty register in the chain. When a read occurs, the contents of the registers shifted one to the right. Control logic keeps track of which registers are empty and which are full. It is typically one extra bit that is loaded with a shifted 0 right on a read and a shifted left 1 on a write.

The size of the FIFO can be statistically or deterministically derived depending on the nature of the traffic into and out of the FIFO. If there is a statistically random set of reads and writes, queuing theory says the average frequency of reads must exceed the average frequency of writes, or the FIFO will require an infinite number of entries to avoid overflowing. If the time between reads

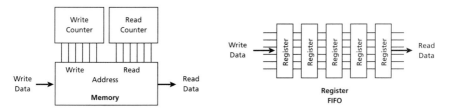

Figure 5.15. Memory- and Register-Based FIFOs

and writes is an exponential distribution, the average length of the queue can be determined, as follows:

1. W equals the mean rate of writes, and R equals the mean rate of reads.
2. If $T = W/R < 1$, the average number of used entries of the FIFO is $L = W/(R - W)$.
3. To calculate the proportion of time that the queue will overflow: Time of overflow $= 1 - (1 - W/R) \star \text{Sum}[(W/R)\star\star K$ for $K = 0$ to the queue length]. $K = L$ gives you the proportion of time the queue will overflow at the average depth.[2]

If the average rate of reads is 8/us, and the average write rate is 4/us, then $L = 4/(8 - 4) = 1$. For a queue size of N, the proportion of time it will overflow is $= 1 - 1/2\star[1 + 1/2...1/2^N] = 1/2^{(N+1)}$.

A queue of 21 will overflow once every $1/2^{22}$ of the time, but the average write is .25 micro seconds or approximately 2^{22} times a second, so on average the queue will overflow once every second. If W/R is close to 1, you need a very large queue to prevent overflows.

The deterministic method simulates the design over a reasonably extreme test case, or calculates the worst case size of the FIFO. For example, if two tasks write into the FIFO 32 words each and then wait until the tasks on the other side have read the 64 words before writing any more, the queue only has to be 64 words deep.

DMA and Bridge Architectures

DMA acts as an agent to transfer information from one target to another. Usually, the information is either in memory and needs to be transferred to an I/O device or the other way around. Typically, there are hardwired lines from a processor to a DMA engine, though a DMA can be a standalone device on the bus. Generally, the processor sends the transaction information to the DMA via a direct connection, and the DMA executes the transaction via a bus transaction.

Some DMAs, like bridges, have enough intermediate storage to get the data from one target and pass it on to the other target, but in other cases it is control logic only. This control logic intervenes between the two targets to coordinate the requests and grants. By requesting a transaction of both devices, one write and one read, it waits until it has both devices and then initiates the transaction. The addresses of the reads and writes are from and to the respective targets. It then coordinates the read data with the write data by changing the address on the bus, so that both devices believe they are transferring information to and from an initiator when they are actually transferring the information directly

---

2. Frederick S. Hillier and Gerald J. Lieberman, *Operations Research,* pp. 404–405.

between themselves. In some systems, this is called "fly by," because the data flies by the device requesting it.

Bridges do not usually employ a fly-by strategy, because the bus widths and speeds are usually different on either side of the bridge. They usually have some intermediate storage to synchronize the data transfer between the two buses. In general, the greater the performance and size differences between the two buses, the larger the intermediate storage needed to insure efficient transfers of data between the two buses.

For example, if the system bus is running at twice the peripheral bus speeds, is four times the size, and one word can be transferred from the system bus per clock cycle, it will take eight clock cycles to transfer that word onto the peripheral bus. The bridge needs at least one word of storage, but must hold off (not acknowledge) the system bus for eight cycles before it can accept the next word. Alternatively, if the bridge usually gets eight word transfers, it can read all eight words into a FIFO, and spend the next 64 clock cycles writing out the burst transfer. Usually the transaction is not complete until all the data has been accepted by the target, regardless of how quickly the bridge can read the data, but some sophisticated buses can split the transaction, that is allow other operations to occur between other devices on the system bus while waiting the 56 cycles it takes for the data to be read by the target on the peripheral bus. For these types of sophisticated buses, the bridge should have FIFOs that are deep enough to handle most bursts of data sent through them from the system bus. In the direction of peripheral to system bus, the data can be collected by the bridge and then be transferred as one burst in a fashion similar to the write.

Flat Versus Hierarchical Bus Structures

It is usually more efficient to have all the devices in the chip on a single bus. There is far less latency in the transfer of data between two devices on a common bus than occurs when the data is transferred between buses. Both buses must be requested and be granted before a transfer can take place between two buses. Unfortunately, in large single bus systems, the more loads a bus has, the slower it operates. If there are a lot of loads on a bus, the resistance-capacitance (RC) delay of the physical structure can affect the performance of the bus. The loads must then be divided among multiple buses, using a hierarchical structure.

The bandwidth limitations of the bus is another reason for building hierarchical structures. When the total bandwidth requests from all the devices on the bus gets above a certain point, for example 40 percent of the bandwidth of the bus, the latency to get access to the bus can be too long. To minimize these long latencies, the devices can be clustered into groups, where each group becomes a bus with far less required bandwidth than the single bus. For example, a bus might be 50 percent utilized by burst transactions that each take 16 cycles to transfer. As was shown in a similar queuing problem earlier, every million or so transactions the latency could get to be 21 times of the average delay, or as much as 336 cycles. In this case, it is more efficient to break up the bus into two buses, ideally with each only slightly above 20 percent. The best place to make that separation is at the point where the least amount of bandwidth is required between the two buses, while keeping the bandwidth of each bus as low as possible.

Using a VC Interface

A VC interface is useful anywhere the designer is connecting two VCs together via their bus interfaces or any VC is being connected to a bus, except where critical timing and gate count considerations require a hand-tailored interface design. The VC interface requires a logic wrapper to interface between the internal logic of the VC or the bus interface logic and the VC interface. Not all of the wrapper logic synthesizes away, and in some cases it is desirable not to eliminate the wrapper logic. The VC interface will become a more well-known and understood interface as tools are developed to support it, but in some circumstances the overhead of the additional wrapper logic might be unacceptable.

For example, the most timing critical areas of an SOC design are typically between the main processor and its memory or critical I/O subsystem. Whenever DMA, or other independent I/O to memory activity, occurs within a system with cache memory, the processor interface must snoop the bus to insure the cache coherency. Timing on a cache miss is critical, because the processor has either stopped or will stop soon. For these reasons, it is desirable to tune the processor interface to the bus in the SOC design, not just for the savings in gates, but more importantly, to save a cycle or two in the transfer of data.

The memory controller is also critical. If the memory is relatively slow, and the controller does not handle intermediate staging of lines of data, the VC interface might not affect the performance that much. However, in cases using fast access memory or multiple threaded access from multiple initiators on the bus, the additional wrapper logic might be too costly. In those cases, tune the memory controller to the specific bus. All other devices will probably not see a large performance hit by keeping the synthesized remains of the wrapper logic or the complete VC interface in the design.

It is expected that even these restrictions will disappear as bus developers and intellectual property (IP) developers create buses and VCs that more naturally fit the VC interface structure. When this happens, the VC interface becomes the natural interface between the bus and the VCs. As such, it will have no significant overhead.

Although a VC interface works with VC to VC connections, it has more overhead than a simpler point-to-point protocol. Even when the natural interface for the VCs is a VC interface, you might need to modify both VCs to eliminate possible redundant clock cycles created by combining both sides of the interface together.

In cases where the VC interface is eliminated by synthesis or was never used because the interface was designed by hand, additional removable logic could be added to create a VC interface for debugging the actual interface. This logic could be removed after the system has been verified, but before the production version masks are cut.

Endian Options

When combining external IP with internal designs, there is always the possibility that the external IP has a different endian type than the rest of the system. The translation from one endian option to the other is context-dependent. You cannot use a fixed translation of the bytes. If the byte addresses between big and little endian were exactly mapped, byte streams would always translate correctly, but half words and full words or larger would each have different translations. Each would require swapping the bytes within the data being transferred.

The bus does not know the type of data being transferred across the bus; only the initiator and targets know. One way to work around this is to have the IP developer produce both big and little endian VCs, and choose the right one for the system. Unfortunately, this is not always possible, especially with legacy designs. However, this option requires no changes to the existing hardware kernels or any other part of the system.

Another option, which is the most common one used today, is to have the initiator keep track of which endian method the various targets are, and map the data accordingly. Frequently, the initiator is a processor, and therefore it would know the context of the data being transferred. With this option, only the software in the processor's I/O handlers needs to be changed, which, unfortunately, can be difficult when a previously verified hardware kernel that does not allow this type of change to its I/O routines is being used.

There is a third option when using an advanced VC interface, which is to transfer the context of the data along with the data itself across the bus. The bus can then be assigned an endian type, and all transactions across the bus would have that endian method. In other words, each VC interface would translate the transactions into and out of the endian type of the bus. In this case, the master and slave VC interfaces on the bus would all have the same endian type. If an initiator and target were both the same endian type as the bus, no translation would take place across those interfaces. However, if a target or initiator had the opposite endian type as the bus, some wrapper logic would be generated as part of the master VC interface wrapper. This logic would look at the size of the data types being transferred to determine what byte swapping needs to be done across the interface.

This option is the least disruptive of the existing hardware kernel. It also enables the derivative developer to connect various endian types to the existing endian type of a hardware kernel. This is the most flexible option, but requires that the data context is built into the VC interfaces and transferred across the bus, which needs to be planned early in the process of creating a platform.

Moving Forward

This section discusses future trends for on-chip communications.

From Buses to Networks

The difference between buses and networks is in the average latency of a read request, as compared to the communication structure's bandwidth. Buses take only a few clock cycles to access the bus and get a response, unless the bus is busy. Networks take many cycles to access a target and get a response back. Sophisticated buses use a large percentage of their available bandwidth, whereas networks begin to suffer if more than 40 percent of their bandwidth is utilized. As we move toward complex SOC design, the size of the chip versus the size of the VCs will grow, leading to relatively slow OCBs. At the same time, the ability to put many wires on a bus means their bandwidths will be extremely large compared to board-level buses today. This suggests that future OCBs will develop more network-like characteristics, and the additional cost of logic will be largely ignored due to the decreasing cost per gate.

From Monitoring to Hardware Monitors

As we move to network-like structures instead of buses, and more programmable elements on the communication structures, the need to tune the arbitration structures becomes important to the performance of the overall system. This will require two things: a way to tune the arbitration, and a way to monitor the transactions across the communication structure to determine what changes are necessary. The tuning requires sophisticated analysis, and will probably be done by the central processor. However, constant monitoring of the system by the processor will burden the processor with overhead. Adding a separate hardware interface that snoops the communication structure and logs the transactions for occasional analysis by the processor might be useful in this situation.

These sophisticated algorithms can employ various learning techniques found in neural network or genetic coding. Future development in this area might result in a system tuning itself for optimal performance, regardless of what application is running.

From VC Interfaces to Parameterizable Interfaces

A major advantage of the VC interface is the parameterizable nature of the interfaces to buses in the future. This will enable many different VCs to be easily integrated on a single bus. The bus side of the interface should contain the bus wrapper parameterization, because there are many more VCs with slave-type interfaces available than buses, and many are legacy VCs. These VCs cannot easily be changed, so the VC interface should be relatively easy to adapt to. On the other hand, the bus interface must be able to translate between many different sizes of VCs, and, therefore, should have most of the parameterization. This puts most of the development burden on the bus developers and VC developers who are making initiators for the buses, but that is far less effort than putting it on the legacy VCs.

There are many parameters to direct the generation of the wrapper, in addition to the actual signals on the VC interface, most of them on the master side of the VC interface. These parameterized wrappers will include sufficient parameters to correctly generate logic for VCs that match the buses they are being connected to, as well as VCs that are very different from the bus. Some of these parameters will be related to the types of transfers that will be done across the VC interface. For example, if only whole words will be transferred, the wrapper on the master side can eliminate the byte enables. In most cases, if a VC with a small interface, such as 8 bits, is connected to a larger bus (64 bits), the bus interface logic can do the assembly and disassembly of the data out of a 64-bit logical word, or only individual byte transfers can occur. A combination of parameters indicating the level differences across the VC interface and application notes defining how extensions should be tied off are required for correctly generating the logic for the VC interface. The master side will at least be capable of connecting to slaves of any level that the master contains.

What's Next

As the industry moves from very large scale integration (VLSI) to SOC designs, another layer of board-level interconnect is integrated onto the chip. About every decade, the level of silicon integration grows beyond the capability of the tools and methodology. Today's level of integration allows for a critical portion of the entire system architecture to reside on the chip. As a result, system communication architectures and hierarchical reusable design methodologies are needed to meet the SOC design challenge. Buses are the dominant structure for system communication. We now need to adapt these technologies to chip design.

6

Developing an Integration Platform

As the industry demands faster time to market, more product flexibility, and more complex designs with lower risks and costs, platform-based design (PBD) provides advantages in meeting these requirements. An integration platform is generally developed to target a range of applications. The breadth of applications targeted depends on trade-offs made in the platform design, such as time to market, cost, chip size, performance, and so on.

This chapter describes the structure of platforms, platform libraries, and the methods for developing and qualifying virtual components (VCs) within those libraries. It also explores the trade-offs that have to be made in developing an integration platform.

In terms of the PBD methodology introduced earlier, this chapter discusses the tasks and areas shaded in Figure 6.1.

Integration Platform Architecture

An integration platform, as discussed in Chapter 3, consists of a library of VCs, a library of embedded software, and one or more hardware kernels for building derivative designs to use within the specified market segment for which the platform was developed. This section provides an overview of the basic architecture of an integration platform.

Hardware Kernels

The hardware kernel is the key difference between a block-based and platform-based design. A hardware kernel must be central to the application being implemented in the chip, because the control of the other VCs in the design resides within the hardware kernel. A hardware kernel needs to contain

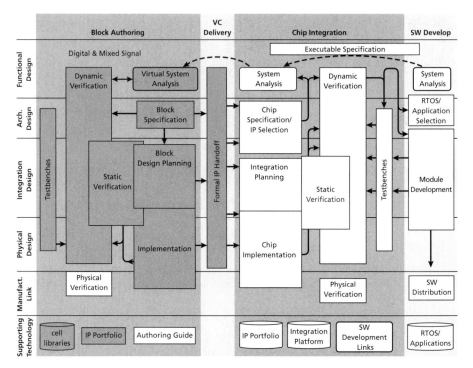

Figure 6.1. PBD Methodology: Developing an Integration Platform

one or more programmable VCs, buses, and VC interface ports. But these basic contents provide no advantage of one platform design over another. To address a specific market segment, additional components, such as a real-time operating system (RTOS), a platform-specific VC, and an interface to programming code, should be included. The hardware kernel should also control the system bus, system power, and test coverage to facilitate quick turnaround of derivative designs.

For instance, an ARM and some on-chip memory, along with an AMBA system bus, might work very well in a set-top box, but would also work well in almost any other application. However, if you add soft modem code, a tuned RTOS, the corresponding residual modem hardware, with enough data storage to handle three HDTV video frames, it would work better for a set-top box application than, for example, a cellular phone.

In addition, an additional VC interface must be available for testing a VC that is connected to a bus, as discussed in Chapter 5, in order to connect the transaction language behavioral model to the derivative design. Eliminating the extra unused VC interfaces after the design has been verified can be part of the physical implementation process.

When creating a derivative, a hardware kernel should not be varied more than is allowed by the hardware kernel's specifications. For example, a hardware kernel can be designed to work with one technology at one clock frequency. If other platform VCs are connected to the hardware kernel to create a derivative, the clock frequency or technology cannot be changed, because there is no guarantee that the design will work.

If one or more elements within a hardware kernel needs to be varied for multiple derivative designs, either the platform should contain different hardware kernels to support each of the derivative designs, or the elements should not be included within the hardware kernel in the first place. For example, if the processor's code size is known, the hardware kernel can contain a PROM. However, if the code size varies widely among the platform applications, a PROM should not be part of the hardware kernel, so that each derivative design can choose one that is appropriate for it.

Conversely, a hardware kernel could be configurable, synchronous, and implemented in a portable physical library such that it can work with a number of different configurations, at clock speeds up to some limit within two or more processes. In this case, a derivative can modify the parameters within the specified limits and still guarantee that the design will work. Elements that are appropriate to make configurable include, but are not limited to, the following:

- Clock speed
- Bus size
- Size of queues within bridges and VC interfaces
- Size of memories (code space, scratch pads, and cache)
- Allocation of address spaces to each of the bus ports
- Structure and priority of the interrupts
- Relative priorities of the initiators on the hardware kernel's internal buses
- Number of technologies and processes the VC will work within
- Amount of dynamic clock and power control
- Availability of various levels of instructions

We will discuss other aspects of configuring hardware kernels in the engineering trade-off section later in this chapter.

Platform Libraries

A platform contains a library of hardware kernel components for creating hardware kernels, and a platform library consisting of one or more hardware kernels and possibly other software and hardware VCs. Figure 6.2 shows the various kinds of elements within the platform libraries.

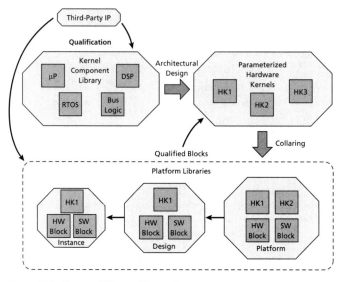

Figure 6.2. Types of Library Elements

To add a VC to the platform library, it must be certified. This means the VC's collar must be modified to guarantee that it will work with all the other VCs within the library. For software VCs, the code must be able to be compiled to execute on at least one of the programmable elements in one or more of the hardware kernels. The software VCs must indicate for which hardware kernels they are certified.

All library components, including third-party intellectual property (IP), must be qualified by running test suites with testbenches before adding them to either the hardware kernel or platform libraries. Qualification can be done by verifying the functionality within a derivative from this platform, as well as by verifying the VC using its own testbench. Parameterized platform VCs are included within a platform library after their degree of parameterization is limited, and the appropriate collars have been added.

Collaring
The different levels of VCs within a platform library are determined by the type of collaring for reusability they have. Table 6.1 lists the attributes associated with the following collaring levels:

- *General* A parameterized VC that is usable in any design that needs the VC's function by specifying the appropriate parameter values. The parameter space is usually too large to verify all combinations for

Table 6.1. Reuse Attributes

Collar Reuse	Functional Verified	Parameterized	Performance Known	Test Known
General	No	Yes	No	Maybe
Platform Specific	Yes	Yes	Maybe	Yes
Design Specific	Yes	Maybe	Yes	Yes
Instance Specific	Yes	No	Yes	Yes

functional correctness. Platform-specific collaring has not been applied at this level.

- *Platform Specific* Similar to general reuse, but the parameter space is limited enough to be able to verify the VC's functionality over the usable parameter space.

- *Design Specific* Design-specific VCs are applicable to a number of implementations of a design, but are specific to a set of derivatives of similar design. The collaring on these VCs is not hardened, though the VC itself might be hardened. The parameter space is limited to performance and encoding of interfaces, but the basic functionality options are predefined. The wrapper is structured for general use for timing, test, and, layout.

- *Instance Specific* Instance-specific VCs have been hardened through a VC design process and can only be used by a specific derivative or revisions of that derivative that does not affect the specific VC. The timing, layout, test, and logical interface have been adjusted and glue logic included to fit in a specific instance of the VC in a specific SOC design.

For the purposes of this chapter, we will deal with design-specific collaring. We are assuming the VCs have not been cast in the target technology's library but have been specified down to register-transfer level (RTL).

Platform Models

Integration platforms have specific modeling requirements that the various components within a platform library need to adhere to. To meet the quick time to market requirements of derivative designs, hardware kernels, in particular, have hard implementations in the platform library. Peripheral hardware VCs need only the instruction set, behavioral, and RTL models, because models of hardware VCs are the same at the architectural and instruction set levels, and they do not have a netlist implementation. Peripheral software VCs need only the architectural, instruction set, and behavioral models. The behavioral level is binary code and does not change through RTL and physical netlist. These assumptions will be used in describing the various characteristics of platform models below.

Hardware kernels require five levels of models, as described below. Hardware kernel component libraries must contain the same type of models as required for a hardware kernel. In some cases, these models are pieces of what will later be integrated into a hardware kernel model, in others they are standalone models. For example, the most abstract level of the RTOS model can include transaction-level communication with application software VCs and the hardware VCs of the design. This is a standalone model that infers the existence of the processor, while a set of binary code for the processor cannot be simulated without the processor model.

Architectural

This functional model of the design consists of behavioral VCs, with no distinguishing characteristics between software or hardware modules, except the assignment of the VCs. It executes the function of the design without regard to timing. This model contains no clock, so scheduling that does occur within the model is event-priority driven. Events are ordered according to the application being simulated. The model is usually written in C or some architectural-level language. I/Os to the testbench are done by passing information through a top-level scheduler. Depending on the application and the architectural models, whole data structures can be passed as data pointed to by the events.

Instruction Set

At this level, instruction set functional models of the processors are added to the design. The event scheduling is brought down to the hardware VC communication level. The software modules are compiled into code for execution on the processors. The memory space for the software applications is added to the design. There is still no clock, and scheduling is done on an event basis, but the data passed is now done at the message level. These models continue to be written in C or some other high-level language. I/Os to the testbench are done by formal message passing through the top-level scheduler. In the process of refining this model, decomposition of the messages down to the packet level can be done.

Behavioral

In the behavioral model, the design is translated into a cycle-approximate model by adding a system clock to the model, and adding delay in each of the functional VCs that approximates the actual cycle delays expected in the implementation of the VC. The scheduler also has latency added to it, so each of the packet-level events have an appropriate level of delay introduced for their transition from one VC to another. The same level of scheduling priority exists as in the upper-level models. This model is usually written in a high-level language that extends down to RTL, such as Verilog or VHDL. The interface between the model and the testbench must also be converted to packet-level transactions, with specific transfer times added.

RTL

At RTL, the design is translated from a scheduling of packets between VCs to a specific set of hardware for the bus and other interblock communication. Registers are added to store and forward the data as necessary. The packet-level data is further decomposed into words and transferred in bursts equivalent to the packet level. This model is an extension of the models above and is, therefore, usually still in Verilog or VHDL. The I/Os have also been translated into specific signals. Data is transferred to and from the testbench on a cycle by cycle basis across these signals.

Physical Netlist

At this level, the design has further been translated down to the gate level. Actual physical delays are added to the design. The timing is now at known increments of time, such as nanoseconds. Signals can arrive and depart at any time within a clock cycle. This model is translated either into a gate-level representation in Verilog or VHDL, or exists within EDIF, an industry-standard gate-level format. The I/Os are individual signals with specific transition times within a clock cycle.

Determining a Platform's Characteristics

Platforms are designed to serve specific market segments. Every market segment has distinct characteristics. We have reduced these characteristics to the relative strengths of the following common factors:

- Performance—from low to high performance requirements of the market place
- Power—from high to low power; low power is more critical
- Size—how critical is the size, or from large to small die requirements
- Flexibility—how much variation is there in the applications, from low to high
- Technology—whether special technology is required, including processing, packaging, IP, voltages
- Reuse—how important is broad use of the platform VCs

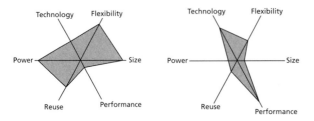

Figure 6.3. Relative Strengths of Specific Factors

Figure 6.3 shows each factor on a separate axis, flattened into two dimensions. The platform applicability is in the area bounded by the irregular closed figure in the center.

The left example might reflect a hand-held wireless device, which needs to be flexible to handle different applications. Small size is critical because of cost. Performance is not critical, but low power is. Technology is moderate, because some analog is needed. The need for reuse is moderately high, because the platform VCs could be reconfigured to serve a number of different applications.

The right example might represent the desktop computing market segment. Technology and performance are critical, but reuse and power are not. Some flexibility is desirable, because these are computing devices, but only a few derivatives are expected, so reuse requirements are low.

Implementing a Hardware Kernel

The hardware kernel component library, which is used to create hardware kernels, consists of processors, memories, buses, software modules, operating systems, and other functional VCs that can interface to the bus. Each component must have relevant models, source code or RTL, and a list of tools and other VCs that they apply to. The component must have applicable tests to verify functionality.

Hardware kernels can be created in the following ways:

- Adding VCs from the platform library to an existing hardware kernel.
- Connecting elements from the hardware kernel component library together to create a kernel.
- Adding one or more hardware kernel library elements to an existing hardware kernel.
- Deleting one or more VCs from an existing hardware kernel.

In the first three cases, the VCs must be qualified to verify that they work within the platform.

To create a hardware kernel from an existing hardware kernel, the original hardware kernel must first be added to the hardware kernel component library, which requires a qualification process. Then the new hardware kernel can be created. This hierarchy allows hardware kernels within other hardware kernels. If access to the hardware kernel's VC interfaces is lost, the hardware kernel ceases to be useful in derivative designs. When using a hardware kernel as a component in another hardware kernel, the parameters available on the component hardware kernel should be set to a specific functional configuration. Another approach is to flatten the component hardware kernel into its subblocks and use them in creating the new hardware kernel. This requires replacement, but can result in the proper ordering of the physical VCs for using the functional parameters of the new hardware kernel.

During the development of the hardware kernel, the timing and area trade-offs have already been made. In most cases, excluding the memory area, a hardware kernel within a derivative should take up more than half the area of the chip. A hardware kernel should have hard implementations to choose from when creating the derivative.

Implementing a hardware kernel requires using a block-based design approach, with additional modules added to address the platform's specific requirements. Some of these modules are described below.

Minimizing Power Requirements

Specific techniques can be used to reduce the power required in the hardware kernel design.

Low-Power Fabrication Process

A number of semiconductor vendors offer lower power fabrication processes, which reduce the leakage and totem pole current to a minimum by shifting the thresholds of the p- and n- channel devices. They also reduce the leakage current by increasing the doping levels of a standard twin tub process. This can reduce the overall power by as much as one half of a normal process, but it also reduces the performance by 10 to 20 percent.

Low-Power Cell Library

A low-power cell library contains all the normal functions, but includes a very low drive. It can be used in small loading conditions. Low-power inverters on the inputs of gates, distributed to the loads of large fan-out nets, can eliminate multiple polarity high-power consuming nets, further reducing power consumption.

Orthogonal Single-Switching Macro Design

Many arithmetic functions are designed in a way that causes some nets to toggle multiple times in a clock cycle. For example, a ripple-carry adder carry line can toggle up to n times per clock cycle for a $2n$-bit adder. Adding additional terms can eliminate this extra switching by selecting the final carry value only, as occurs in a carry-select adder.

Active Clocking

There are two levels of active clocking. The first uses clock enables. The power consumed in any CMOS design is proportional to the clock frequency. Within a complex SOC design, not all units need to operate at the same time. To conserve power, those units that are not active can be "shut off" using clock enables. Current design practice in full scan designs is to use data gating. In this mode, the data is recirculated through the flip-flop until it is enabled for new data. This takes far more power than gating the clock, and the further back in the clock distribution structure the gating takes place, the less power is consumed in the clock structure. When using this technique, care must be taken to not create timing violations on the clock enables. The enable signal on data-gated flip-flops has the same timing constraints as the data signal itself. Unfortunately, the enable signal for clock gating must be stable during the active portion of the clock, which requires that more stringent timing constraints be met by the enable signal.

The second level of active clocking is clock frequency control, in which the original clock frequency of selected clocks is slowed down. Many parallel operations are not used or complete earlier than they need to in SOC design, but because they must periodically poll for interrupts, they must continue to operate. In these cases, the operations could be slowed down, rather than disabled by controlling the VC's clock frequency. The hardware kernel's clock control unit should have independent frequency controls for each of the clocks being distributed to different subblocks in the hardware kernel. There are many design considerations, including timing the design for the worst case combinations of frequencies between units. Care must be taken to insure that the function is invariant over all such combinations, which might require additional holding registers to ensure that the data is always available for the next VC, which might be running at a slower frequency.

Low-Voltage Design

The best way to reduce the power in a chip is to lower the voltage, since power is a function of the voltage squared, or $P = 1/2CV^2$. For designs that are not latency-sensitive, the voltage can be lowered. This slows down the logic, requiring a slower clock. If the bandwidth must be maintained, additional pipeline stages can be put in the design to reduce the amount of logic between each stage, resulting in the same bandwidth and clock frequency, in a substantially lower power design. This alone can reduce the power below half of the original chip's consumption. Additional design must be done between the internal logic and the chip outputs, since the signals must be level shifted to avoid excessive I/O current.

Active Power Management
Units that do not need to be active can be turned off completely, thus eliminating the power loss during the off period. This is done by dropping the power rail to ground for that unit only. Care must be taken to maintain any critical states in non-volatile storage and to hold outputs at legal switching levels to minimize totem pole current in the adjacent devices. Special scheduling and reset circuitry must also be added, since it can take a large number of clock cycles to bring up a VC that has been turned off.

Maximizing Performance
Several steps can be taken to increase the performance of a hardware kernel.

Translating Flip-Flops to Latches
Unlike flip-flops, latches allow data to pass through during the active portion of the clock cycle. Latches allow the slow, long paths in a cycle to overlap with the shorter paths in the next cycle. Figure 6.4 shows that the same logic between flip-flops and latches is 25 percent faster in the latch path because of cycle sharing.

Staggered Clocks
In early microprocessors, it was a common practice to stagger the clocking of registers to take into account the delays between the first and last bits in the arithmetic function being executed. Typically, the first bit of any arithmetic operation is a short logical path. In an adder, it is a pair of exclusive ORs; in a counter, it is an inverter. On the other hand, the last bit often takes the longest. In an adder and a counter, the carry path is the longest path and the last stage is in the highest bit, as shown in Figure 6.5.

In this example, the clock can be delayed by adding a pair of inverters between each bit in the registers, starting at the lowest bit. The staggered string should not be extended for too many levels, because the clock skew can exceed

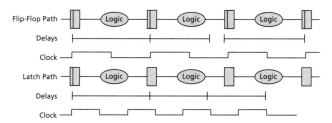

Figure 6.4. Cycle Compression by Translating Flip-Flops to Latches

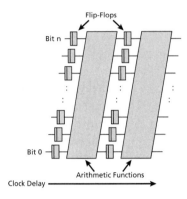

Figure 6.5. Staggered Clocks

the advantages of the skew. This is also more useful if the arithmetic functions are iterative, since the data must be unskewed at I/Os, because buses and memories are not usually skewed.

High-Performance Design

Much of the detailed design implementation method is built into timing-driven design methodology. This drives the synthesis of the design by the clock cycle constraints of the paths, which serves to level the critical paths in the design. If the design is not latency-sensitive, but requires high bandwidth, additional registers can be added to the pipeline in the design, reducing the amount of logic and corresponding delay between the registers in the design.

High-Performance Architecture

Steps can be done at the architectural level to create high-performance designs. The first is to select a bus that has the latency and bandwidth characteristics necessary to meet the design's performance objectives. The second is to duplicate resources, as necessary, to parallel what would otherwise be serial operations. The third is more complicated. It requires reorganizing operations that require serialization into parallel tasks by eliminating the serializing constraints. For example, a serialized summing of numbers requires each number be added to the total. This function can be modified so that successive pairs of numbers are added together, with their corresponding sums added until the whole set of numbers is totaled. This restructuring enables the second step of duplicating resources to be executed. Refinements to this process can result in multiport operations, such as memories, three port adders, and so on.

Minimizing Size Requirements

Certain tasks can reduce the physical size of the hardware kernel.

Design Serialization

This process is the reverse of what is done in high-performance architectures. Area is usually saved whenever duplicate resources can be combined by serializing the resource. However, the wiring should not be increased so much as to offset the gate savings.

Tristates and Wired Logic

In general, a set of multiplexor (MUX)-based interconnects is much faster than tristates, but almost always takes up more area. One way to reduce size is to replace any MUX-based switching structures with slower tristate-based bus structures. However, the bus logic should be timed so as to minimize the periods when the bus is floating or in contention.

A second way to reduce area is to convert highly distributed fan-in/fan-out structures into wired logic. For example, an error signal might come from every VC in the design. These must be ORed together to form a system error signal that must be broadcast back to all the VCs. For n VCs in the design, this structure would require n input signals, n input or gate, and an output with n loads. This can be translated into a wired OR structure that has one wire and an I/O from every VC, as shown in Figure 6.6. In this case, the size reduction is in the wire savings.

Low Gate Count Design

This is the reverse of the low-power or high-performance design. To reduce area, slower smaller implementations of arithmetic functions can be used. For example, use a ripple-carry adder instead of a carry select, or a serial–addition multiplier instead of a booth's encoded fast multiplier.

Low Gate Count Implementation

This is the reverse of high-performance design in that the synthesis should be done to minimize gates, and pipelining should only be used where absolutely needed for performance. Similarly, powering trees should be kept to a minimum.

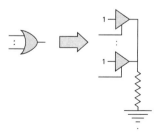

Figure 6.6. Tristates and Wired Logic

Maximizing Flexibility

Flexibility can be achieved in two areas: in the amount of parameterization that allows for greater configurability of the hardware kernel; and in the use of programming to allow more variety of applications in the hardware kernel.

Configuration

Soft configuration or parameterization can be used to configure a system. Soft configurations include configuration registers in the design to select the various function options. These options should be set during bring-up and changed via RTOS commands from the processor.

Parameterization can take on two different forms. It can be used to drive logic generators, or it can be used to select options that already exist within the design. Both parameterization approaches should be defined by key word, but the latter form means the key words translate into tie-off conditions for the selection logic. Hard VCs can only be configured by this approach. All options that are parameterized must be qualified. At a minimum, a hardware kernel should have parameterized VC interfaces. The VC interface parameterization should be at least as broad as the capabilities of the bus to which it is connected.

Design Merging

When two similar designs are merged, they can include two or more functions and MUX-selecting between them. Design merging can also generalize two or more dedicated designs by abstracting their common datapath and developing merged state machines that execute each merged function based on the value of the configuration bits. The parameterization of this approach is similar to soft configuration, where the configuration signals are driven by flip-flops rather than tie-offs. In another mode, the configuration bits could be hard-wired, as in the parameterization option described in the "Configuration" section above. If the VC is not yet hardened, synthesis should be run to eliminate the unused logic cut off by the configuration state. In this case, the merged VC should be designed to maximize the elimination of unused logic, which can be done by minimizing the amount of unused logic that is driven by and drives flip-flops in the design.

Programmable Elements

Programmable elements can include everything from traditional field programmable gate array (FPGA) logic to processors. The general control processor provides greater flexibility and ease of modification, while the FPGA provides more performance for a more limited range of applications. This is because execution of applications in the processor is mostly serial, while execution of the applications in the FPGA is mostly parallel. Also a traditional

FPGA's configuration is only loaded once at each system reset, while the processor can load different applications as needed. Digital signal processors (DSP) and other special purpose processors, as well as dynamically reconfigurable FPGAs, fall in between these two extremes. The selection of the type of programmable elements is done at the architecture phase of hardware kernel development.

Hardware to Software Mapping

A hardware kernel is more flexible if it allows more programmable options. To the degree that VCs are included in the hardware kernel, they may be implemented as application programs rather than hardware VCs. This selection process is done at the architecture phase of developing a hardware kernel.

Timing Requirements

The hardware kernel must have accurate timing models at the physical level, except for the additional soft collar logic, which can be timed during integration. The peripheral blocks need only models accurate to the clock cycle and some estimated delay for the I/O paths. When implementing the hardware kernel, the external peripheral block designs must have sufficient delay to enable the hardware kernel's internal bus to function properly.

Typically, the hardware kernel is hardened with soft logic for the collar. If the unused ports to the internal buses can be removed, the timing models for the hardware kernel should include parameters for the number of used ports, since the upper bound of the operating frequency of the bus is related to its loading.

Clocking Requirements

The hardware kernel must generate the clocks required for a derivative design and distribute those clocks to the other VCs in the design. The hardware kernel must clock the peripheral VCs because the devices need to be clocked in relationship to the processor and system bus, and the frequency control has to be available beyond the processor for the derivative to work.

In the implementation of a derivative design, the hardware kernel takes up half the chip, with uncontrollable blockages on all routing layers. At the top level of a derivative design, the routing from one side of the hardware kernel to the other is a problem. Where the external blocks will connect with the hardware kernel is known, and in most derivative designs, all the other VCs connect to the hardware kernel. Therefore, providing the clock to the VCs from the hardware kernel is much easier and safer from a skew consideration than trying to match the skew from a global clock. Since the hardware kernel is so much larger than the other VCs, routing the clock around it would require a lot of padding on the clock to the peripheral blocks. This would increase the skew well above a flat clock distribution system, so a clock system distributed through the hardware kernel makes the most sense.

The clocking requirements for the peripheral block designs do not require a built-in clock structure, because they are mostly in RTL form. Therefore, the clock structure within a hardware kernel should fan out to the VC interface ports first, and then branch to the specific regions of the VC. This provides an early clock to further minimize the clock skew between the peripheral components.

Transaction Requirements

A hardware kernel must have at least two VC interfaces. The bus within the hardware kernel must be designed to handle all or none of the VC interfaces. Every VC with a VC interface must have some compliance test suites, which are written in a relocatable form of the transaction language for the VC interface, to verify the underlying VC. The VC interfaces on the hardware kernels must be flexible enough to handle the interface requirements of most of the peripheral blocks in the platform library.

All peripheral VCs should be designed with sufficient buffers and stall capability so that they operate in an efficient manner if the bus is not at the same level of performance as the datapath. Regardless of which frequency the system and any specific peripheral block is operating at, the design must still function correctly.

Physical Requirements

A hardware kernel must be implemented in hard or firm form, and the collar should be soft. The hardware kernel should target an independent set of multi-foundry libraries or have an adjustable GDSII file available. If the hardware kernel is firm, the clock structure should be pre-routed according to the clocking requirements. The test logic and I/O buffering logic in the collar should be initially soft, to allow some flexibility during place and route. Space should be allocated for this collar, including the possibility of sizing the I/O buffers to reduce the delay between VCs.

Test Requirements

Each block, including the hardware kernel, should have sufficient observability and controllability so that it can be tested in an efficient manner. Test logic must be included to be able to isolate and apply test cases that verify the design on a block by block basis. The collar should include the boundary scan for the block as well as the access method for the test logic within the block. This logic should be structured to allow easy integration with other VCs and the test access port at the chip level.

Hardware kernels with processors should have some additional test mechanisms to allow loading and using any internal memory within the hardware kernel or external memory through a VC interface and to execute diagnostic programs to check out the processor, the system bus, and as much of the operation of the peripheral VC as is reasonably possible. This on–chip diagnostic capability enables large portions of the hardware kernel to be tested near or at speed. These types of tests can detect timing-related faults, as well as structural faults, that are normally caught with scan-based approaches. In addition, some method for directly accessing the system bus, in conjunction with the boundary scan of the peripheral VCs, can be used to test the functionality of the peripheral VCs. This is especially useful if the peripheral VC has no internal scan test capability.

Power Requirements

Each hard VC's power and ground distribution should be structured to do the following:

- Distribute the maximum power the VC can use
- Not exceed the metal migration limits of the process
- Not exceed one-fifth of the allowable voltage drop for the entire power distribution structure
- Connect to external power and ground rings easily

Hardware kernels might require two levels of power distribution, because of the wide variations in maximum power requirements for the different sections of the design. This two-level structure should have major wires between each of the subblocks, and some grid-like structures within each subblock section of the design. Verification of the power distribution structure can be done with commercially available power simulation tools.

Instance-specific VCs include power and ground rings around them, because the VC's position within the overall chip and the chip's power grid structure are known.

Software Requirements

All software modules must be associated with one or more operating systems and processors. Hardware kernels must include references to one or more operating systems that have been qualified to run on those hardware kernels. The operating systems must include drivers for the devices that exist on the hardware kernel as well as shell drivers for the interfaces. The operating system requirements also include being able to handle (either through parameterization or in the actual tables) all of the interrupt levels available in the processor.

The application software modules must be either in source or in relocatable binary object code. Test cases should exist to verify the functional correctness of the software module. If the software is in source, it must be certified to compile correctly. If the software is in object code, it must be certified to properly load via a defined link and loader. Ideally, the software should use a standard interface definition for passing parameters. This enables creating a parameterizable top-level module that can be configured to fit any application by iteratively calling the underlying software modules. In the absence of such a top-level module, an example application or testbench could be reconfigured to create the derivative designs. Hardware kernels must also have some diagnostic software test suites to verify the proper function of the programmable elements within the VC.

Verification Requirements

Much of the verification requirements of a derivative design can be obtained from the model levels described earlier. Additional testbenches are required for the VCs, ideally in the transaction language, so they can be reused in the integrated derivative design. Hardware emulation, or rapid prototyping, is a commonly used verification technique, but it requires an additional model view.

If rapid prototyping is used in the derivative design, building a board-level version of the hardware kernel, with the bus at the board level connecting to the VCs that reside in separate chips as shown in Figure 6.7, is the most efficient way to do the emulation. Peripheral designs in FPGAs can then be connected to the board-level bus. When this is done, both sides of the VC interface reside within the FPGAs. This requires mapping from the existing hardware kernel design to one that does not contain any VC interfaces. The hardware kernel's bus is routed around the rapid prototyping board to each of the FPGAs. Each of the FPGAs has the VC interface to bus wrapper in parameterized form. When the derivative design is emulated, the peripheral VCs' VC interfaces are connected to the FPGA VC interfaces.

Additional ports on the board-level bus can be used to access the whole design for observing and initializing the design. If the peripheral components are memory VCs that exceed the capacity of the FPGA, the FPGA will just contain the translation from the bus to the memory protocol, with the memory external to the FPGA. Some facility for this possibility should be designed into the hardware kernel rapid prototype. It can be a daughter card bus connector assembly, or an unpopulated SRAM array on the main board. SRAM is preferred in this case, because the emulation of other types of memory can be done in the FPGA interface, but SRAM cannot be emulated by the slower forms of memory.

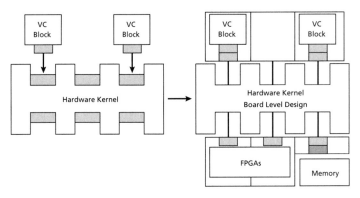

Figure 6.7. Rapid Prototype Mapping

Engineering Trade-offs

This section explores how to decide when to employ the various techniques used to design VCs and hardware kernels.

Performance Trade-offs

Blocks in the platform library should be designed to allow the widest range of clock frequencies possible. This is easier to achieve with the smaller soft peripheral VCs, because synthesis can tune the VC for the desired performance. The hardware kernel must be over-designed, because it should be hardened for quick integration into derivative designs. In general, these designs should be synchronous with as much levelization of the long paths as possible. It is more important to design the hardware kernel to the required performance levels needed in the derivative designs than the peripheral VCs, because the performance of the derivative design is determined by the performance of the hardware kernel. The peripheral VCs also often have less stringent performance requirements.

The system bus should be over-designed to provide more than three times the bandwidth than the expected configurations require. The reason for this is that the hardware kernel should be designed with more VC interface ports than needed in the anticipated derivatives, in case they might be needed. If the extra VC interfaces get used in a derivative, more bandwidth will be required. The additional loading reduces the top frequency the design can operate at, which has the effect of lowering the bandwidth of the bus.

The processor should also be over-designed, or the interfaces to the bus should be designed to allow for wait states on the bus, because the most lightly loaded bus configuration might be able to run at a frequency above the normal operating range of the processor. If the clock frequencies are configurable within the VC, the wait states can be employed. Otherwise the hardware kernel's performance is bound by the processor, not the bus.

The memory configuration is a critical part of the system's performance. Different applications require different amounts of memory. Usually, it is better to keep larger memory requirements outside of the hardware kernel, using cache and intermediate scratch pad memory to deal with fast iteration of data. Unfortunately, memory that is external to the hardware kernel usually requires more clock cycles to access than memory that is internal to the hardware kernel. In general, slower memory, such as DRAM or flash, should be external to the hardware kernel, whereas limited SRAM and cache should be internal. The bus interfaces to the memory that is internal to the hardware kernel should be tuned to eliminate any extra clock cycles of latency to improve the system performance.

Every hardware kernel should have at least one function that is specific to the platform that the hardware kernel was designed for. If it is a hardware VC, it is usually some number crunching datapath design, which is replacing software that would do the same job, only slower.

Given the variety of technology options available to the user, make sure that the hardware kernel operates correctly. Different semiconductor processes create different performance variations between memory and logic. DRAM processes provide space- and performance-efficient DRAMs, but the logic is much slower than in a standard CMOS process. All the devices that reside on the bus should be designed to stall, and the bus should be able to allow the introduction of wait states to ensure that the hardware kernel functions correctly, regardless of the technology used to implement it.

Sizing Trade-offs

A number of architectural decisions can significantly affect the size of a derivative design. Size can be most significantly affected by whether the proper amount of the right type of memories are used in the design. It is best to leave the larger slower forms of memory outside the hardware kernel, because they will need to be tailored to the specific application's requirements.

It is more cost-effective if a hardware kernel can be applied to as many derivative designs as possible. One way this can be accomplished is to over-design the hardware kernel with all the memory, processors, and special components that the derivatives for its platform would need. Unfortunately, the hardware kernel would be too big for most of the derivative applications. On the other hand, it is impossible to use a hardware kernel that has less than the customer's required options. It is, therefore, more size-efficient to create a number of options either through parameterization or by creating a family of hardware kernels, some of which have limited options, such as internal memory size, bus size, number of processors, and number of special components.

Then, the chip integration engineer can choose the hardware kernel that is closest to but greater than what is required.

Designing Hardware Kernels for Multiple Use

The best solution for addressing size issues is to use a combination of parameterization and multiple hardware kernels. The more parameterized the hardware kernel is, the larger the amount of logic that must be soft, to either synthesize away the unused portions or drive the generators to build the hardware kernel without the unwanted options. This results in more implementation work during the platform integration of the derivative design. If the hardware kernel is mostly hard, the excess parameterization results in poor performance or excessive size. The large number of options also creates a problem, since all the options must be verified to qualify the design. If the verification is not done, there is an increased risk that the derivative design will not work, thus increasing the verification and debug effort.

If there is too little parameterization, it is more likely that the hardware kernel would not meet the derivative's requirements, resulting in additional design work to modify the hardware kernel to meet the derivative's requirements. If the hardware kernels are not parameterized, more of them are required to efficiently cover the platform's application space, which is more up-front work for the platform developer.

A trade-off must be made between the degree of parameterization and the hardware kernel's coverage of the platform's design space. The ideal solution is to design the underlying parameterization of the hardware kernel as generally as possible and limit the options available to the user.

Since the hardware kernel has a lot of options, one or more will fit the peripheral VC's interface, thus reducing the verification requirements for the peripheral VCs. New hardware kernels can be quickly implemented, because the only work is to create a new user option and verify it. A trade-off exists here as well: although it is usually less work to verify a peripheral VC than a hardware kernel, there are many more peripheral VCs than hardware kernels in the platform library. Lastly, general parameterization is often easier to design than specific options, because the underlying hardware kernel components can be used, it can be done locally, and the logic can be hierarchical.

For example, an interface can be optionally added to the hardware kernel at one level, while a FIFO can be optionally inserted in the interface locally at a lower level. The local option could have been previously defined as part of the parameterization of the interface itself. User options are then preconfigured sets of values for all the general parameters defined in the design. To provide sufficient flexibility while limiting the options, the external options should be broken into two types: functional and interface.

Functional options add or subtract specific functional capabilities in the hardware kernel design. The hardware kernel is implemented with these functional capabilities as hard subblocks within the hardware kernel itself. The options can be eliminated as necessary, without having to be soft. These options should be limited to subblocks, preferably on the physical corners of the hardware kernel, to minimize the raggedness of the resulting rectilinear shape, although this limits the number of reasonable options.

Interface options are implemented in the soft collar, and can be more flexible without incurring the size penalty of hard subblock elimination. The interface logic can be verified locally, since its scope is local and only needs to be verified for one instance, although many might be present in the design. For example, a hardware kernel has many copies of the VC interface, yet only one instance of each type, master, and slave needs to be qualified, providing they all connect to a common bus in the same manner. Also, if the hardware kernel's VC interface has sufficient options that cover all the interfaces of the peripheral VCs in the platform library, parameterization of the VC interfaces of the peripheral VCs is not necessary. To make this trade-off, use the following process:

1. Set $K = 1$.
2. Sort all of the peripheral VCs by their VC interface options.
3. Eliminate all VC interface options used by only K or less peripheral VCs. The remaining options define the parameterization set for the VC interfaces in the hardware kernel. The peripheral VCs eliminated must have their VC interfaces parameterized to meet at least one of the VC interface options in the hardware kernel.
4. Estimate the amount of qualification necessary for the peripheral VCs and hardware kernels.
5. If the cost of peripheral qualification > cost of hardware kernel qualification, increment K, and go to step 2.

A similar process can be used to determine how many hardware kernels must be created versus how many options each VC should have.

Incremental Versus Initial Library Construction

While a platform can be created from an abstract analysis of the requirements of a market segment, the risk is much lower if specific designs within that market segment are analyzed. The common elements in these designs form the basis for hardware kernel definitions, and the rest of the elements form the initial basis for peripheral VC definitions. If a hardware kernel library does not exist, the elements defined in the hardware kernel definitions form the basic elements for that library as well. Work can and should begin on both the component and platform libraries, but pragmatic business requirements often result in contracting derivative designs that require VCs that are not in the library. After concluding that the derivative design does indeed belong within this platform, the derivative's requirements further define the elements that must be added to the platform library. In this way, the platform library is incrementally expanded to cover the intended market segment.

Moving Forward

The notion of an integration platform is just beginning to be developed today. At the board level, some platform designs have already been used. The decision to create a motherboard delineates the board level hardware kernel from the peripherals. Still, using this approach is just beginning at the chip level. In this section, we look beyond the simple, fixed, single processor-based notion of a hardware kernel and explore the types of parameterization and programmability that could be created for a hardware kernel.

Parameterizing VCs

Many of the functions of a hardware kernel can be parameterized, ideally separate from the interface protocol. Because all the elements of a hardware kernel are optional to some degree, the physical hardware kernel should be organized to allow for the most likely options; that is the most likely options should be on the corners and edges of the VC. Other rules to consider include:

- Memory used exclusively by an optional processor or other VC should be outside that VC.
- The bus should partition the design in a regular fashion if the bus size is expandable.
- Protocol options should be soft if possible, or require little logic to implement.
- Memory should be expandable on the hardware kernel's edge or corner.

For example, if a hardware kernel contains a scratch pad memory that is solely used by a co-processor within the design, the memory should be outside of the processor or at least occupy an equivalent amount of the hardware kernel's edge. An expandable bus can have the number of address and data lines increased or decreased. The bus must span the entire VC in the vertical and/or the horizontal direction, so that the movement of the subblocks does not create gaps with respect to the other subblocks in the design.

Some of the protocol examples include: VC interfaces definition, which should be soft configurable wrappers, implemented in the collar; interrupt lines and priorities, which can be hard-coded and tied-off, or implemented in the collar; and arbitration logic, which, if centralized, should reside on an edge, otherwise it should be part of the collar.

Figure 6.8 shows a preferred placement of a parameterized hardware kernel before and after parameterized options have been applied. In the before view, all the options are in a standard configuration. The interrupt and VC interface logic is soft and in the collar of the VC. The scratch memory is connected to the co-processor. The buffer memory is associated with the application-specific block, and the I/O devices are ordered so the optional one is on the outside.

The original hardware kernel is rectangular, but by eliminating the co-processor, its memory, a VC interface port, and an I/O option, along with doubling the bus size, adding cache, and reducing buffered memory, the shape changes to a rectilinear block. In the original, the bus had two branches, but one was eliminated. This is acceptable, because the aligned, rectangular blocks can still be packed without creating wasted space. If the VC interfaces are hardened, eliminating one of them would create an indent in the resulting hardware kernel, which if small enough, would become wasted space. Options resulting in small changes in the external interfaces of a hardware kernel are better done as soft collar logic, unless the operation of the logic is so timing critical that it requires a specific physical structure.

This example is not necessarily the best or the only way of organizing the hardware kernel. If the buffer and scratch memories were combined across the

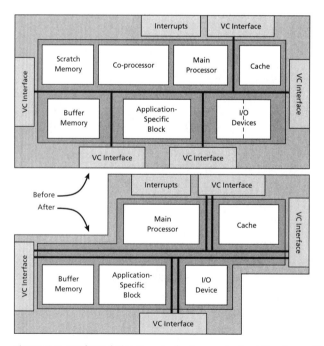

Figure 6.8. Preferred Placement of a Parameterized Hardware Kernel

whole edge of the hardware kernel, and the I/O and co-processor were arranged on the other edge, the result might have remained almost rectangular, but performance limitations might prohibit such organization. Still, it is a good idea to include the limitations of the placement of the hardware kernel when analyzing how much parameterization the VC should have. Usually, the more rectangular the resulting shape of the hardware kernel is, the less likely there will be wasted space in the derivative designs that use it.

Configurable Platforms

In the future, platforms will contain one or all of these programmable or configurable structures:

- Classic stored programs that serially configure a processor that executes them
- Reconfigurable logic, including FPGAs
- Soft, configurable, semi-dedicated structures, which we call configurable functions

Reconfigurable Logic

Depending on the speed and frequency of reconfiguration, reconfigurable logic can have many different implementations, some of which are discussed below.

Slow Reconfiguration

Slow reconfiguration, at 10s of milliseconds, should be used only on bring-up. This is similar to the existing reprogrammable FPGAs today, such as the older versions of the Xilinx 4000 series. This configuration is useful as a prototyping vehicle and is generally loaded serially from an external ROM or PROM.

Fast Reconfiguration

Use fast reconfiguration whenever a new function is required. At 10s of microseconds, it is fast enough to have a number of configurations in external memory and load them when a new type of operation is requested by the user. The operation itself is still completely contained within the programmable logic for as long as it is needed. For example, the Xilinx 6200 series has a programming interface that looks like an SRAM, and it has the capability to load only parts of its configurable space at a time.

Instant Reconfiguration

Use instant configuration, which is 10s of nanoseconds, during the execution of a function, as required. Logic can be cached like programming, and parts of a process can be loaded as needed. In this case, the bus must contend with a significant bandwidth requirement from the reconfigurable logic. However, this

type of hardware is flexibly configurable to handle any operation. This ultimately leads to compiling software into the appropriate CPU code and reconfigurable hardware, which can later be mapped into real hardware, if necessary.
Examples of this can be found in the current FPGA literature.[1]

Configurable Functions

Configurable functions are usually programmed by setting configuration registers: I/O devices are often designed to be configured to the specific type of
devices they communicate with; arbitration and interrupt logic can be configured for different priorities; and a clock control system can be configured for
different operating frequencies. Here, we will focus on functions that can be
configured during the device's operation, as opposed to hardwiring the configurations into the devices before or after synthesis. The latter approach was
discussed in the parameterization sections earlier.

Figure 6.9 shows the relationship of the different types of programmability
to performance. The left graph, which assumes an equivalent on-chip area,
shows the relationship between flexibility and performance. That is, the number and sizes of applications that can be executed versus the speed at which the
application can execute after initial loading. A stored program is the most flexible, since it can handle any size application and do any function without the
need for complex partitioning. For the various forms of reconfigurable logic,
the speed of reconfiguration relates to the ease with which reconfiguration can
be considered part of the normal execution of an application. Instant reconfiguration means the application can be broken into many smaller pieces, thus
providing more flexibility, but since it is logic, it provides more performance
than a stored program. Slow reconfiguration limits the flexibility partially

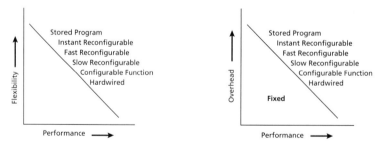

Figure 6.9. Flexibility and Overhead in Relation to Performance

1. Steve Trimberger, "Scheduling Designs into a Time Multiplexed FPGA," *International Symposium on Field
Programmable Gate Arrays,* February 1998; Jeremy Brown, et al., "DELTA: Prototype for a first-generation
dynamically programmable gate array," *Transit Note 112,* MIT, 1995; and Andre DeHon, "DPGA-coupled
microprocessors: Commodity ICs for the early 21st century," *IEEE Custom Integrated Circuits Conference,* 1995.

because of the slow reconfiguration, but has somewhat less overhead, so the size of the logic for the equivalent area is larger than instant reconfigurable and hence performs better than the instant reconfiguration, which must swap more often. Configurable functions and hardwired functions are severely limited in flexibility, but have less performance overhead, so are faster.

The right graph shows the relationship between the amount of silicon area dedicated to the programmability versus the performance of the functions that can be run on the different options. The hardwired approach has almost no overhead; it might have some reset logic, but little else. The configurable functions have some registers, whereas hardwired would only have wires. The reconfigurable options must have programming logic, which is serial for the slow reconfigurable function and parallel in the fast option. For the instant reconfigurable, the programming logic must also manage multiple cache line-like copies of the configurations, but the processor that executes the stored program must be considered almost entirely overhead, except for the execution unit, for managing the programming.

Figure 6.10 shows the relationship between the cost and performance for each type of programmability. The left graph shows the relationship of the application overhead to performance, which can be viewed as the area per usable logic function in the application. Since the stored program's processor is fixed in size and can handle, via caching, any size application, it has little application overhead internal to the programmable VC. At the other extreme, hardwired logic is all application, and every addition of functionality requires more logic. The configurable function is somewhat less because a number of options are combined, so the area is less than the sum of the separate functions, but they cannot all be run in parallel as in the hardwired case. Slow reconfiguration requires a larger array to effectively run large applications. Instant reconfiguration has less overhead for the application, so it can more effectively reuse the programmable logic.

The graph on the right compares the cost of storing the programming. Stored programs use more memory as the application grows, but most of the

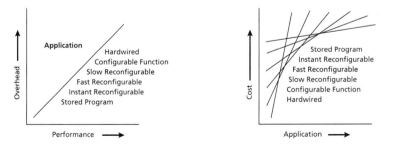

Figure 6.10. Relationship of Cost and Performance

cost is in the processor, so the slope of the top line is nearly flat. Each of the other options requires more hardwired logic as the application grows, down to totally hardwired. Instant reconfiguration is incrementally more cost-effective than slow reconfiguration, because the program or logic maps can be stored in less expensive memory outside of the FPGA logic, whereas the slow must contain the application.

Note that these graphs overly simplify a relatively complex comparison. Specific cases might not line up as nicely, and the scales might not be exactly linear. For example, FPGAs typically require 20 to 50 times as much area per usable gate than hardwired logic, so an application that requires 1,000 gates is one-twentieth the cost in hardwired logic. The curve does not appear as steep as suggested in the last graph above, but considering that far more than ten functions can be loaded into the FPGA VC, the effective cost for the gates in the FPGA could be less than in the hardwired case. Similarly, the spacing of the various programmable options on the graphs is not linear as implied on the graphs. In fact, the difference in performance between instant and slow reconfigurable logic, not including loading of the configurations, is very similar and considerably faster than the stored program, because the execution is in parallel rather than serial. On the other hand, the hardwired and configurable function options are very similar in performance, but as much as two to five times faster than the reconfigurable options.

The graph in Figure 6.11 is probably a better picture of the complex relationship between the performance and functionality of the various devices. The relative performances might be off, but the relationships are correct. This graph assumes that each structure has the same silicon area, not including the external program space, and that the initial configuration was loaded at bring-up.

The hardwired has the highest performance, but can only execute when the application can fit in the area. The same is true for the configurable function,

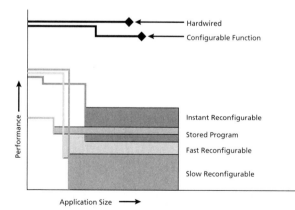

Figure 6.11. Relationship between Performance and Functionality

but the dip in that line is due to the cost of reconfiguring. The same is true for all the reconfigurables. Each has a dip in performance when the application size exceeds the capacity of the configuration space. The second dip in the instant reconfigurable is when all its internal configuration spaces are exceeded. The stored program has a dip when the application size exceeds its cache size. The broad bands for performance indicate the variation of performance the devices might have. The processor might have more or less cache misses depending on the code. The instant reconfigurable might execute out of its cache for sufficient time to reload, or it might experience a large cache miss and stall waiting to load the next configuration. The fast and slow reconfigurables will stall, but for how long relative to the execution depends on the application.

This leads to some conclusions about the trade-off between different types of programmability. In general, well-known small applications with high performance requirements should be hardwired. Similar applications with specific variations and/or options that might not be known at the time the chip is created should be cast as configurable functions. This is one option of the configurable function; the other is to hardwire the options if they are not needed during execution of the design.

At the other extreme, if there is a need to run a wide variety of applications of different sizes, some of which are quite large, and they contain a very high amount of control logic branching, the preferred solution is the stored program processor. RISC machines should be chosen for predominately control operations, while DSP could be chosen for more number-crunching applications. If more performance is required, the size and number of applications are more limited, and the functions are more data-manipulation oriented, a variety of levels of reconfigurable logic should be chosen. The more the application can be broken into small self-contained functions that do not require high performance when swapping between them, the lower the speed of reconfiguration needed. For example, for a multifunctioned hand-held device that has GPS, modem and voice communications, and data compaction, but each one is invoked by a key stroke (100s of milliseconds), the slow reconfigurable might be the most appropriate. But if a hardwired solution was needed because of performance, but the applications vary widely and have a wide variety of control versus data requirements, the instant reconfigurable option might be the only one that can do the job.

In the future, hardware kernels will have some mixture of these configurable options. Thorough analysis is necessary to determine how much, if any, is appropriate for any specific hardware kernel. In the next chapter, we continue this discussion from the perspective of implementing a derivative design.

7

Creating Derivative Designs

As discussed in the previous chapter, the key to platform integration is the existence of a platform library containing qualified, collared virtual components (VC). This chapter describes methods for successfully creating and verifying a derivative design. In terms of platform-based design (PBD), this chapter addresses the tasks and areas shaded in Figure 7.1.

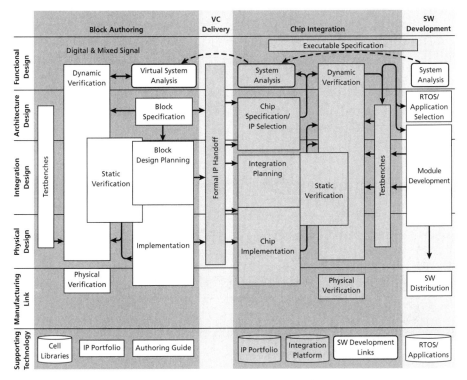

Figure 7.1. PBD Methodology: Implementing a Derivative Design

The Design Process

To create a derivative design, a specification is needed, which can require significant system-level simulation prior to implementing the derivative design. The technical requirements for the design, such as the package, pin out, and external electrical requirements, must also be defined. Then the design must be mapped to an existing set of VCs within an existing platform library that encompasses the defined technical requirements; otherwise, the selection of VCs from the library will not meet the derivative requirements. The end result is a top-level netlist that specifies the platform VCs to be used in the derivative design.

The platform library contains all of the necessary models and block-level testbenches, but a top-level testbench must be created before or during the mapping process to verify the correct implementation of the design. This implementation produces a set of test vectors and the mask data necessary to fabricate and assemble the derivative design. The testbenches and models can be modified to create a rapid prototype to verify the design before implementation, and to debug the system after the derivative chip is integrated.

Figure 7.2 is an example of the structure of a derivative design. The hardware kernel, which is enclosed in the dark line, contains VC interfaces in the soft collar (the area surrounding the hardware kernel). These are connected to the internal bus (multiple-grouped lines), which is distributed in the hard block (shaded in dark grey). The subblocks are arranged within the hardware kernel. Around the outside of the hardware kernel are peripheral VCs with their collars and their corresponding VC interface wrappers. The clock structure is driven by an analog phase locked loop (PLL). Together these make up the derivative design. Not all the VC interfaces on the hardware kernel are used (see top of diagram), and other interfaces besides the VC interface can have soft collar logic as well (interrupts).

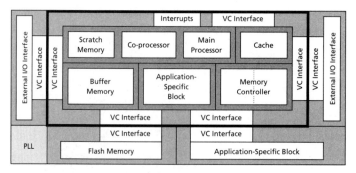

Figure 7.2. Example of a Derivative Design

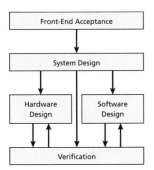

Figure 7.3. Design Process

Creating a derivative design involves the phases shown in Figure 7.3. Front-end acceptance is similar to the process in block-based design addressed earlier. During system design, the architectural design is created and mapped to a set of platform VCs. Hardware design includes design planning, block design, and the chip assembly processes. In software design, the software components are designed and verified. Verification is an ongoing process throughout the development of the derivative design. All of these processes are discussed in more detail in this chapter.

Front-End Acceptance

Front-end acceptance is the process of reviewing the requirements specified by the user or customer of the design, and estimating the likelihood of meeting those requirements by following the platform integration process. Some of the analysis, such as determining whether the performance and power requirements can be met or whether packages exist for the pin or silicon requirements, can be answered based on knowledge of the available technology and experience with similar designs. Some of the critical requirements might require further analysis or "dipping."

Dipping, as shown in the flow chart in Figure 7.4, involves doing some of the tasks of system, hardware, or software design on a specific set of requirements to ensure that the derivative design can be constructed. The tasks can include some of the mapping and hardware implementation of a subsection of the design. Generally, though, as little as is necessary should be done in this phase.

If the design does not meet the specifications, the integrator returns to the customer requirements phase to obtain revised requirements for creating the derivative design.

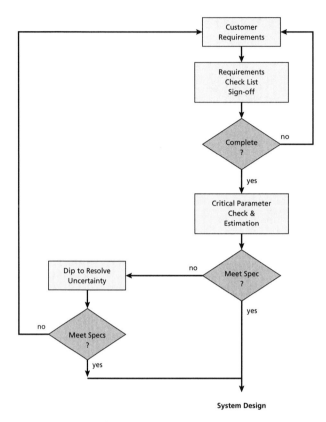

Figure 7.4. Flow of Front-End Acceptance

Selecting a Platform and VCs

After it is determined that the design can be created, the system design phase begins. This step includes refining the system's requirements into an architecture. This step, function–architecture co–design, starts with abstract functional and algorithmic models, high–level views of an SOC architecture, a mapping of the behavior to that architecture, and performance analysis in order to validate and refine the design at a high level. This process can be facilitated with performance modeling tools.

After the appropriate architecture is determined, the integrator must map the architecture onto a platform and select the VCs within that platform to implement the derivative design. Depending on the platform level selected, many of the VC and architecture choices might have already occurred as part of the platform definition. Platforms are defined to serve a specific market segment, so the integrator must select those platforms that cover the market segment served by the target design. There might be many platforms that cover the market segment.

Platform and Derivative Characteristics
Some of the derivative design's requirements can be analyzed using the six characteristics shown in Figure 7.5. Five of these characteristics and how they apply to platform VCs were described in the previous chapter. The sixth, time to market (TTM), replaces reuse, since the derivative design is not being implemented for reuse, although it is related to TTM. The primary measure of TTM is the schedule to create the derivative design, which is primarily a function of the complexity of the design and the applicability of the VCs within the platform library. The more generally reusable the VCs are, the more work that must be done at integration, so reuse and TTM are inversely related. A missing factor is the derivative's cost, because it is a function of all the other factors. The higher the performance, power, technology, flexibility, size, or TTM requirements, the higher the cost of the resulting derivative. Therefore, cost can be viewed as the area within the shapes shown in Figure 7.5, given the proper scaling of each of the characteristics.

To select the proper platform from those with the same market segment, compare the platform's characteristics as described in the previous chapter, with the derivative's requirements shown here. When comparing Figure 7.5 to the diagrams in the previous chapter, the requirements are similar, except that TTM and size do not match. A high TTM requirement corresponds to a low reuse requirement for the hardware kernels in the platform. The size requirement of a derivative corresponds to its expected size, while the size factor for a platform corresponds to how critical size is.

Selecting a Hardware Kernel
When selecting a hardware kernel, all of the requirements should be evaluated. For example, in the diagram to the right in Figure 7.5, the derivative has high performance, large size, and short TTM requirements for the application, as compared to lower power, small size, and lower performance requirements with moderate TTM requirements on the left. The size of each hardware kernel, its

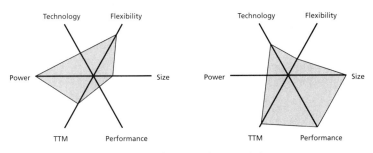

Figure 7.5. Relative Strengths of Design Characteristics

performance, and how close it fits to the derivative's requirements should be compared. Keep in mind that a poor fitting hardware kernel for the diagram to the right is worse in performance, but it might well fit the requirements of the diagram to the left if the size is not too large. The hardware kernel that meets the goals is the best, which can be determined by creating a weighted sum of the hardware kernel's capabilities, where the weights are a translation of the derivative's requirements. For example, the diagram to the right has a short TTM, large size, and a high performance requirement, so these parameters are heavily weighted; the low requirements for power reduction and technology receive low weightings. The subsequent rating then is:

$$\text{Rating} = \text{Sum}(\text{Derivative's weight}*\text{hardware kernel's capability})$$

System-level performance modeling tools provide an efficient means to explore the effectiveness of a number of chosen hardware kernels by mapping the application to them and comparing the system-level response.

Selecting Peripheral VCs

A similar process is used to select the peripheral VCs after the hardware kernel is mapped to the derivative's architecture. The VCs in the platform need to be sorted by their ability to fill a requirement in the mapped derivative design and to connect to the chosen hardware kernel. Because more than one alternative for each possible peripheral VC could exist, or one VC could fill more than one function, begin with the function that is covered by the fewest VCs in the platform library. Rate the peripheral VCs using the weighted sum above, and pick the one with the highest rating. After picking that VC, other peripheral VCs with overlapping functions could be eliminated. Repeat this process until all the VCs have been chosen.

Again, using system-level tools allows various peripherals to be explored and to determine the impact at the system level.

Integrating a Platform

Top-Level Netlist

Mapping the architecture to the selected VCs is usually done at the architecture level, where the implementation of modules in hardware or software is not distinguished. To assign the modules to their proper VCs, they are first mapped to the instruction-model level based on the software/hardware mapping, and then mapped down to the behavioral-model level. At this level, the platform VC's models can replace the derivative's modules by adding glue logic to match the interface requirements. Eliminating the glue where redundant yields a top-level netlist.

Verification and Refinement

Up to this point, the derivative design has been simulated with a testbench derived from the original system requirements, and mapped down to the behavioral level. Further refinement is done by substituting the register-transfer level (RTL) models for the equivalent blocks in the behavioral model. In some cases, a number of parameters might need to be set to generate the appropriate RTL for the behavioral block. This should be done in accordance with the requirements of the derivative design, which partially translate into the requirements specified for each block at the behavioral level.

The hardware kernel's cycle-accurate model can be used to verify the functionality and performance at this level.

With the proper assertions at the behavioral level and below, formal verification techniques can also be used to help verify the successive refinements of the design. Together, the two verification techniques minimize the verification cycles—minor changes to the design are formally verified, leaving major regression cycles to the traditional simulation approaches. Most formal verification programs can compare RTL to netlist-level designs for functional equivalence, without the need for an assertion file. The assertion files can be used in place of behavioral to RTL formal verification, which might not be robust enough.

VC Interfaces

After the top-level netlist and a cycle-accurate simulation model exist, the VC interface parameters are derived. In Figure 7.2, the VC interfaces on the hardware kernel connect to the system bus, and the VC interfaces on the peripheral VCs connect directly to the peripheral VC. The VC interface should have the same data and address sizes as either the peripheral VC or the system bus, whichever is possible given the range of parameterization of the VCs. Usually, more parameterization (flexibility) is available on the hardware kernel, so the VC interface size that is safest to use is the one that matches the peripheral VC's internal address and data bus sizes. If that is not possible, use the same sizes on both sides of the interface and add interconnect logic to match the VC to the system bus size.

Clocks

At this point, the clocks are hooked up by connecting the available clocks on each hardware kernel VC interface to the clock pin(s) on the peripheral VCs. At this time, the PLLs are hooked up between the chip's clock pin and the hardware kernel. The hardware kernel contains the clock distribution for each of the clocks in the design. The required delay of the clock trees in the peripheral VCs is specified for each of the clock outputs from the hardware kernel. These should be applied to their respective peripheral clocks. In some cases, the hardware kernel might have multiple asynchronous clocks. This can still be managed in timing analysis on a clock by clock basis if the following rules apply:

- The hardware kernel distributes these clocks to separate ports.
- The peripheral VCs use only clocks that are synchronous to each other.
- No direct connections exist between peripheral VCs that have clocks asynchronous to each other.
- The hardware kernel deals with all asynchronous communication that occurs between logic that uses the asynchronous clocks completely within the hardware kernel.

Verifying that no direct connections exist between peripheral VCs that have clocks asynchronous to each other can be done by simulating to stability after setting all internal states in the design (to avoid unknowns) and setting each of the clocks to unknown by repeating this process for each clock. After each clock is unknown, check the interconnects between the peripheral VCs on VCs that have a known clock value. If any of the interconnects have unknown values, this rule has been violated.

I/O and AMS VCs

Usually, the I/Os and analog/mixed-signal (AMS) VCs are placed on the periphery of the chip. These VCs contain their own pads, power, and ground rings. A single VC might contain one or more pads. These pads should be spaced according to the requirements of the semiconductor vendor. Special voltages are either provided as part of the power and ground rings (since the I/O power and ground is separate from the internal power and ground) or as part of the I/O pads of the cell. The internal power and ground may not be part of the original block, unless it is used within the block, but in either case it is included as part of the collar for the VCs.

As shown in Figure 7.6, the collar should extend the power and ground rings to the edges of the cell to allow for a butting connection with the portion of the rings created in the power module. As such, the power and ground at the edges of the VCs match the requirements to be specified in the power module. To insert an I/O cell, the existing power and ground rings are cut and replaced. Only digital signals using the internal power and ground voltages are valid connections between these VCs and the rest of the internal logic in the chip. As a result, all level shifting and conversion from analog to digital is done within the I/O or AMS VCs.

The package selected must have enough physical pins for all of the design's signals, test logic, and power and ground pin requirements. The silicon vendor can provide the power and ground I/O requirements.

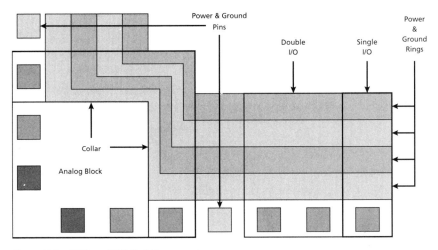

Figure 7.6. AMS and I/O Cell Placement

When placing I/O blocks, the locations of fixed power and ground pins on the selected package, and the estimate of the additional power and ground pins required to minimize power and ground bounce, must be taken into consideration. The pin assignment might have to be altered following the power distribution design to allow for additional or replaced power and ground pins.

The I/O blocks are placed so that powers or grounds can be added at the end of each row. Simultaneously switching I/Os are placed near the dedicated power and ground pins of the package, if possible. This reduces the need for additional power and ground pins. Either organize the pins to minimize the wiring congestion on the board, or organize the I/O VCs to minimize the routing requirements for the chip, while keeping the logical groups of I/O contiguous for easy usage. Position AMS VCs along with the I/O on the edge of the chip. As much as possible, place AMS VCs next to dedicated power and ground pins to reduce the signal coupling between the AMS signals and the digital signals.

With the rise of special high-speed I/O structures, like RAMBUS, the distinction between traditional I/O and analog is blurring, which is why I/O cells and AMS blocks have been grouped together in this analysis.

Test Structures

The test structures are planned before the VCs are implemented. Testing a complex SOC consists of a hierarchical, heterogeneous set of strategies and structures. A mixture of traditional functional tests, scan-based tests, built-in self-test (BIST), and at-speed testing should be used, as necessary. Traditional scan is useful for catching structural flaws in the chips, but it is not adequate for verifying the chip's performance. This requires at-speed testing. Integrated circuit designs and processes are becoming so complex that traditional methods of guaranteeing SOC performance using timing libraries are inadequate for ensuring that all structurally sound devices from wafers that meet the process limits will also meet the performance requirements of the design. This is partially due to the increasing inaccuracies of the timing models and the increasing variation of individual process parameters as we move into ever deeper submicron geometries.

At-speed testing is one solution to this problem. Much of the hardware kernel could be tested with at-speed functional tests if the internal cache or scratch pad memory can be used to run diagnostic tests. This approach, like BIST, could alleviate the need for high-speed testers in manufacturing. A low-speed or scan-based tester loads the diagnostic or BIST controls. A high-speed clock then runs the BIST or diagnostic tests and unloads the results via the slow tester clock. This approach only requires generating a high-speed clock through special logic on the probe card and is far less expensive than a high-speed tester.

Even if scan is not used internally within the VCs in the design, use it to isolate and test the VCs individually within the chip. This enables each of the VC's tests from the platform library to be used, which greatly reduces the test generation time. Unfortunately, traditional IEEE 1149.1 joint test action group (JTAG) control structures do not have facilities for multilevel, hierarchical test construction. Additional user-specific instructions must be added to enable the JTAG at the chip level to control all the levels of scan and BIST strings in the chip. Each of the blocks or VCs at the top level of the design should have their own pseudo JTAG controller, which we refer to as a VC test controller (VCTC), as shown in Figure 7.7. A VCTC controls other VCTCs internal to the block. The hardware kernel's VCTC needs to connect to the VCTCs of the subblocks within the hardware kernel. At the lowest level, the VCTC controls the BIST or scan-based testing of the VC.

At this point in the design process, the structure of the physical test logic is defined. The top-level JTAG controller logic and pins are added to the top-level netlist, along with the connections to the blocks. The block-level test logic is added in VC design, and the test generation and concatenation occurs in chip assembly.

Power Requirements

Power requirements are translated into power rings around the blocks, and an interconnect structure is defined to distribute power to the blocks from the top chip-level rings around the edge of the chip. The diagram in Figure 7.8 shows how such a structure is wired.

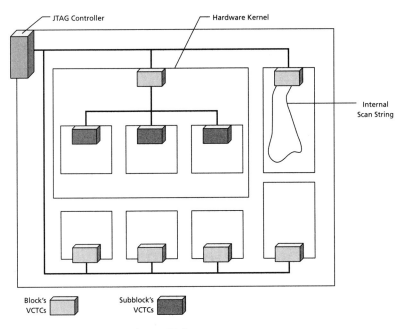

Figure 7.7. Test Structure for an SOC

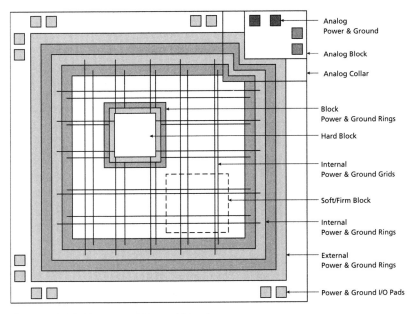

Figure 7.8. SOC Power and Ground Structure

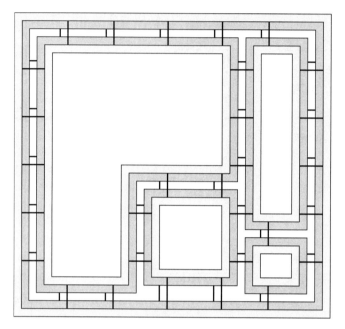

Figure 7.9. Power and Ground Distribution System

Connecting all the blocks in the design yields a power and ground distribution system shown in Figure 7.9. The hardware kernel, which takes up almost half the chip area, is on the left.

This structure might seem redundant when compared to a single set of power and ground stripes between the blocks in the design, but it has a number of advantages:

- It is block-oriented, so blocks can shift at the top level without changing the block requirements for their rings.
- The underlying grid is easier to calculate than other structures, and is thus easier to plan.
- The global interconnect is properly external to the block, so changes in routing do not affect the block layout.

Floor Planning

Lastly, the models are migrated to the logical netlist level. Since the hardware kernel has a hard implementation, the detailed timing of its I/Os is available. Any functional parameters must be set at this time to create the proper footprint for the hardware kernel. Detailed timing is then done to verify that the design meets the requirements. To reduce the slack, negative slack is used to drive the floor planning, which consists of manually placing the hardware kernel and the peripheral blocks initially near their respective VC interfaces. These are then moved as necessary to meet the design's timing and size constraints.

Block Implementation

During block implementation, the most hardened level of each block that also meets the specified constraints is selected and the collar logic is created by applying the specified parameters to the generators or manually designing it. The clock tree is created using the specified delay obtained from the hardware kernel. In addition, the test logic is included in the collar logic according to the specification from chip planning. The soft blocks are placed and routed, and the hard blocks' collars are placed and routed to the size and shape specified by the floor plan. The hardened blocks should have shapes that are readily placed in most derivative designs for that platform, so no repeated layout is required.

Chip Assembly

In this step, the final placement, routing, test pattern generation, and integration are done. Then a GDSII physical layout file is created, and the appropriate electrical rules checks (ERC) and design rules checks (DRC) are performed. The test vectors are generated, converted into tester format, and verified as well. The information necessary for a semiconductor vendor to fabricate the chip is created.

Throughout this process, the various transformations of the design must be verified using the techniques described in the section on verification below.

Analyzing Performance

Throughout the process of design implementation, the performance is analyzed. At the architectural level, timing does not exist within the simulation, but the known bandwidth requirements can be translated into some level of performance estimates. For example, an architectural-level model of an MPEG2 decoder might not have any timing in it, but the total number of bytes created between modules for a given test case can be counted. The test case is translated into a specific maximum execution time assuming the video requirements of 30 frames per second. At the instruction level, timing is also not explicit, but again, the number of instructions can be translated into a specific processor that is known to have a specific MIPS rate. At the behavioral level, estimated time in approximate clock cycles is now explicit, which translates into a minimum required clock frequency. At RTL and netlist level, static timing analysis is done in addition to simulation.

With the clock frequency defined, static timing analysis at RTL provides a rough estimate of the timing (most algorithms use quick synthesis). This can be used to drive the synthesis tool to insure that the minimum clock frequency does not produce violations in the design. At the netlist level, logical timing is more accurate than RTL timing, but physical back-annotated timing is better: usually within 5 to 10 percent of the SPICE simulation at the transistor level.

In platform integration, the static timing analysis that is done is focused on the top-level netlists. Hardened blocks provide their physical back-annotated timing, while soft netlist level blocks use their logical timing. Floor planning and block-level synthesis are driven by these static timing analysis results to generate better design performance.

Verification and Prototyping Techniques

A platform-based design needs to be functionally verified. This section describes the process of migrating a testbench through successively more detailed models to guarantee that the design works at every level. Functional simulation is very time-consuming, especially when simulating the more detailed models. The techniques for rapid prototyping described here reduce the time required to verify the derivative design.

Models for Functional Verification

Testbench migration is required to guarantee functional equivalence from the highest to the lowest levels. As the testbench migrates from one level to the next, it is further refined to meet the requirements. Refinement must include additional test suites to test functions that are not directly checked by the converted test suites. This section describes the testbench requirements for each of the model levels: architectural, instruction set, behavioral, RTL, and physical netlist.

The key to successfully verifying each level of successive refinement is to start with the previous level inputs and verify the functionality of the model over the aspects of the test suite that are invariant between the two levels. At the same time, the level of detail and the number of points of invariance should increase as the successive refinement is done.

Functional Level
At the functional level, the testbench uses interactive or batch file input that represents the full system's input. The output is also translated into a form that is equivalent to the full system's output. For example, a testbench for an MPEG decoder would consist of inputting one or more MPEG video files, and output would be one or more pixel files that can be viewed on a monitor. The testbench translates the I/O into this human recognizable form. In batch mode, it must handle the file access, or handle the I/O devices for interactive applications.

Application Level

The testbench at the application level tests the software code. In addition to the functional level I/Os, there is code that must be loaded initially and, depending on the application, before each test suite is executed. The code either resides in a model of memory in the testbench or in a memory block in the model. The data that is transferred to and from the model must also be broken down into messages. These are larger than a packet, but smaller than a whole file or test case.

The point of invariance from the functional level is primarily at the output file equivalence level, where each functional-level I/O file constitutes a test suite. Tests to check the basic operation of functionally transparent components, such as the reset and boot operations and cache functions, if present in the model, are also added.

Cycle-Approximate Level

At this level, the testbench is refined from message-level to packet-level transfers to and from the model. A clock and time is added as well. The testbench from the application level must now gate the data at a rate for the clock in the model that is equivalent to the real-world operation. Output is also captured on a clock basis.

The points of invariance from the application level are primarily at the message equivalence level and the program/data memory states at the original points of functional invariance. At this level, appropriate latencies within and between blocks in the design are also verified. Typically, these tests are diagnostics that run in a simplified manner across limited sets of blocks in the design.

Cycle-Accurate Level

The cycle-accurate testbench is refined from packet-level transfers to word-level transfers to and from the model. Data is now captured on a clock-cycle basis, where a packet of data might take many cycles to transfer. The data is also decomposed into individual signals from the higher level functional packets. Output is created on a cycle by cycle, single-signal basis.

The points of invariance from the cycle-approximate level are at the packet-equivalence level and selected register and program/data memory states at the original points of application invariance. Adjustments to the testbench output interpretation or test suite timing must be made to insure these states are invariant. This might require either reordering the output to insure invariance or choosing a subset of the message-level points. Tests to check the proper functioning of critical interactions in the system, such as cache to processor, the bus operations, or critical I/O functions, are also added.

Timing-Accurate Level

The testbench is refined from word-level transfers to timing-accurate signal transfers to and from the model. Every set of signals, including the clock, switches at a specific time, usually with respect to the clock edge. Output can be captured in relation to the clock, but it is transferred back to the testbench at a specific point in the simulation time.

The points of invariance from the cycle-accurate level are at the clock-equivalence level, and most register and program/data memory states at the original points of cycle-approximate invariance. Adjustments to the testbench output interpretation or test suite timing must be made to insure these states are invariant. This might require either reordering the output signals between adjacent clocks to insure invariance or choosing a subset of the packet-level points. Additional tests at this level include checking the proper functioning of critical intraclock operations, such as late registering of bus grants, cache coherency timing, and critical I/O functions that occur within a clock cycle.

Example of the Testbench Migration Process

In this example, the functional level model is a token-based behavioral model. The cycle-approximate behavioral model has a complete signal-specific, top-level netlist. To verify the resulting netlist with the original, token-level test vectors, additional layers to the original testbench must be added.

The original testbench provides tokens, which are data blocks of any size, into and out of the model. It controls the transfer of the data based on specific signals, not clocks, since clocks do not exist in the behavioral model. The behavioral blocks are targeted for either hardware or software, depending on the resulting architecture that is chosen. A simplified diagram of this is shown in Figure 7.10.

To use the same token-level tests, it is necessary to translate from the token-level to the cycle-approximate level. Since the token level has blocks of data that could be any size, two levels of translation must occur. First, the data must be broken into packets or blocks that are small enough to be transferred on a bus. Second, the proper protocols must be introduced for the types of interfaces the design will have. Figure 7.11 shows this transformation.

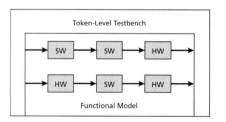

Figure 7.10. Initial Testbench Migration Model

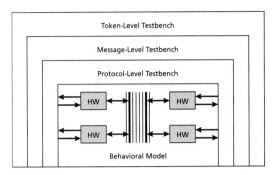

Figure 7.11. Example of Testbench Transformation Hierarchy

In this example, the cycle-approximate behavioral model has a correct signal-level netlist that connects to hardware VCs only. A clock and an interface for external or internal memory for the object code from the original software VCs in the functional model must be added. The specific pin interfaces for the types of buses the chip will connect to on the board also must be added. The procedure described here for accomplishing this requires standard interfaces that are part of the Virtual Socket Interface (VSI) Alliance's on-chip bus (OCB) VC interface specification.

In the behavioral-level model, the types of interfaces are defined. The easiest way to simulate using those protocols is to use in the testbench a similar block opposite to the interface block that is used in the chip. The signal pins can then be connected in a reasonably easy way. Next, the VC interface on that interface block is connected to a behavioral model that takes in messages and writes them out as multiple patterns on the VC interface. Lastly, another program that reads in tokens and breaks them down to message-level packets to transfer to the VC interface-level driver is added. The diagram in Figure 7.12 depicts this translation.

Now the top-level token to message translator all the way to the VC interface can be done using the OCB transaction language. A series of routines converts token-level reads and writes into VC interface calls. The chip interface block is a mirror image of the interface block inside the chip. For example, if

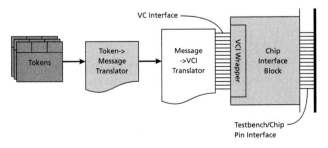

Figure 7.12. Structure of Testbench Transformation

the chip has a PCI interface on it and a target device internal to the chip, the testbench should contain a behavioral model of a initiator PCI interface. These mirror image protocol translators are defined as follows:

Mirror function -> Interface function = 1 or no changes from VC interface to VC interface (up to the maximum capability of the specific interface used)

The token to message translator must be at the transaction language level described in Chapter 5. Some code must be written to translate tokens into a series of transaction calls, but the rest can be obtained from either a standard translation technique or from the platform library (in the case of the chip interface blocks).

Verifying Peripheral-Hardware Kernel Interconnect

Vectors for testing each of the VCs are in the VSI Alliance's OCB transaction language form in the platform library. The VCs are individually tested with their vectors first, and then later a testbench is used to communicate from the transaction language through the bus block to the individual blocks.

Each VC in the system is verified. These same vectors are then interleaved in the same fashion that would be seen in the system. This is an initial system-bus verification tool, which later can be augmented with system-level transactions. The transaction language has three levels-the highest is VC–interface independent, while the lowest is VC–interface, cycle-timing specific. The middle transaction language is a series of relocatable VC interface reads, writes, waits, and nops. The test vector files contain the specific VC interface signal values as parameters, along with variables for relocatability. Reads contain the expected information, and writes contain the data to be written. Assignment statements and global variables allow the creation of test suites that are address- and option-code relocatable for use in different systems. This feature results in a new methodology for migrating the testbench and applying it to successively more accurate models, while keeping the same functional stimulus. This technique is shown in Figure 7.13.

Each peripheral VC has its own test set, previously verified via the stand-alone VC testing technique. The hardware kernel has at least two VC interfaces for any specific bus; one is used for each of the target blocks and the other to connect the behavioral VC interface model. If all of the VC interface slots are assigned to peripheral VCs, the peripheral VCs should be removed to free a slot for the behavioral VC interface model. The vectors for the resident peripheral VCs are then executed through the hardware kernel model's bus.

The system vectors need to cover at least the common characteristics among all the devices in the platform library that can be connected to the hardware

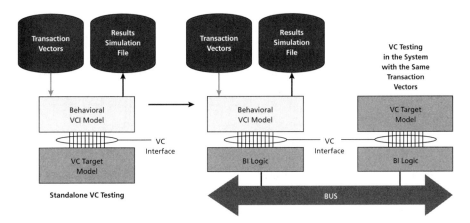

Figure 7.13. Standalone Versus In-System Testing

kernel, probably in the form of an I/O software shell or control routine. This can be run on the hardware kernel as an additional hardware kernel-based system diagnostic. The diagnostic does not need to use the transaction language to generate or observe bus-level transactions. It can contain only functional vectors that can be observed on the external pins of the hardware kernel. Either the pins go directly to the chip's I/O pins or a special testbench must be created to access these pins.

Verifying and Debugging Software

To verify and debug the software, instruction-set simulators and cross compilers can be used to verify the functionality of the software modules before loading them into the prototype systems. Special debug compilations provide tracing and check-pointing as well. Other special compilations accumulate statistics about the execution of the code. These profilers produce tables that show where the code was most executed or the size of the numeric values that were computed. These tools are useful in optimizing the code for better performance.

Some of these capabilities are extended to rapid prototyping facilities as well as the target chip by providing test logic inside the processor, which returns the low-order bits of the current instruction address in real time to the external pins. This capability can be included in the hardware kernel, using the JTAG interface as the external interface. Software that emulates the operation based on a map of the actual instruction space also exists. It interprets the instruction address and emulates the operation of the instruction in parallel with the hardware, so that the software designer can have complete visibility of the software module's code as it executes on the processor.

Rapid-Prototyping Options

The typical derivative design size is large enough to require many millions of clock cycles of testing. This creates regression simulations that can run weeks or months when using an RTL or gate-level simulator. Verification then becomes the bottleneck when developing a derivative. Using emulators or programmable prototypes, which are only one order of magnitude slower than the real device, reduces verification time.

For example, Quickturn's hardware emulators, which are massive groups of Xilinx, SRAM-programmable field-programmable gate arrays (FPGA), are configured so that large logic functions can be mapped onto them. Aptix's or Simutech's quick prototyping systems have slots for standard devices, buses, interconnect chips, and FPGAs that enable more targeted emulation.

The most appropriate rapid-prototyping device for platform designs consists of a hardwired hardware kernel and slots for memory or FPGAs off of the hardware kernel's bus. The hardware kernel contains a model with its VC interface removed to allow mapping to such a hardware prototype. The peripheral VCs are synthesized into the FPGA logic and merged with the VC interface logic from the hardware kernel. To create a suitable model for rapid prototyping, the process of chip planning through block implementation must target the FPGA only.

The primary I/Os go either to a mock-up of the real system, or to a tester-like interface that can control the timing of the data being applied to the rapid prototype and retrieved from it. In the former case, special clocking and first-in first-out (FIFO) storage might need to be added to synchronize the slower prototype to the real-time requirements of the mock-up. For example, in a rapid prototype of an MPEG2 set-top box, the signals from the cable or satellite must be stored and transferred to the prototype at a much slower rate. This can be done by having another computer system drive the video data onto the coax-input cable at a slower rate. The television set on the other side must have either a tape for storing the screens or at least have a single-frame buffer to repeat the transmission of each frame until the next frame arrives. If not, the image will be unrecognizable by the designer.

Experience has shown that creating this type of model often has its own lengthy debug cycle, which reduces the effectiveness of rapid prototyping. One way to avoid this is to debug the hardware kernel rapid prototype with a reference design before making it available for the derivative design debug.

The debug process should include the ability to view the instruction execution, as well as monitor the bus traffic through a separate VC interface. This is done easily by connecting the bus to some of the external pins of one of the unused FPGAs. This FPGA is then loaded with a hardware monitor that transfers the data back to the user's computer through a standard PC interface, such as a parallel I/O port or USB. The traces are used to simulate the RTL or lower models up to the point of failure to obtain internal states in the design, which are not viewable through the hardware.

Breaking up the test suite into smaller and smaller units to isolate the bug without using a lot of simulation time minimizes the debug effort. One way to do this is to create a standard bring-up or reset module. Then organize the test suites into segments, any one of which can be executed after applying the reset sequence, regardless of the previous tests that were applied. For example, the MPEG2 test suites can have limited sequences between I frames. If the test suites are broken into groups beginning with an I frame, the test suite segment that failed can be run after the reset sequence on a simulation of the design in a reasonable amount of time. Furthermore, if the failing cycle or transaction is identified from the traces off the rapid proto-type, the data capturing on the simulator is done on the cycles of interest only, thus saving even more time, because the I/O often takes more time than the simulation itself.

Engineering Trade-offs

Some of the engineering trade-offs regarding designing an integration plat-form include selecting the VCs to use in a design, and implementing and ver-ifying a platform-based design.

Selecting VCs

When comparing parameterized hardware kernels with non-parameterized ones, what the parameterized hardware kernel would reduce to needs to be estimated. If the specific parameters are not yet determined, the TTM for the parameterized kernel is worse than for the non-parameterized one, because the parameterized kernel must be generated with the correct para-meter settings. If the correct parameter values are known, they are applied to the parameterized hardware kernel. This produces a design that is equivalent to the non-parameterized one for better comparison. If more than one viable option from a parameterized hardware kernel is available, ideally all options are created. Unfortunately, this can result in too many options, which in this case only the extreme options are created. When there is a continuous range for a parameter, use the results of the two extreme cases to determine the correct setting for the derivative design. Sometimes this value can be calculated. For example, a bridge might have a parameterized depth for its FIFO. Both no FIFO and the largest FIFO can be created, but it might be easier to look at the traffic pattern and just create the correctly sized FIFO. If not, a simulation with monitors on the FIFO would obtain the same results.

When comparing a hardware kernel that contains a number of the features needed in the derivative design to a hardware kernel with more VC interfaces and fewer options, selecting which is the better one depends on whether the peripheral VCs can cover the same features as the hardware kernel. If the less flexible hardware kernel does not have an exact fit, use the hardware kernel with more VC interfaces, which is more flexible, unless the debug costs are excessive. It is important to factor in the time required to debug the additional features compared to starting with a prequalified hardware kernel.

When choosing VCs for a derivative design, equivalent functions in both software and hardware form might exist. If both options meet the required performance and the MIPS exist in the processor for the new VC, use the software solution. However, it is more likely that something must be added to the hardware to fit this block into software. In that case, the VC should go in hardware if the design has not reached its size limits. If when mapping the design, a VC wants to go into hardware, but the only VC in the platform library that contains the desired function is in software, the function must be put in software, and the design is mapped from that point to determine what other functions can be put into hardware in its place. Most of these decisions are made when designing an integration platform, but iterations might be required between choosing the platform and VCs and mapping the rest of the system.

Reconfiguring vs. Redesigning

Ideally, most parameterization is either in the soft collar of a hard VC or is in soft VCs, which means the parameters are set and the resulting logic can still be optimized through synthesis, regardless of the type of parameterization. If the VC is hard, the choices are either tying option pins to specific values or using the options provided in configuration registers. Configuration registers provide more debug capability. If the wrong options are selected prior to implementation, they can still be modified at system bring-up. Generally, building options into the chip, which can be configured out during bring-up rather than leaving them out, is safer, especially when the difference is a small amount of logic. This is because system debug often turns up situations that were unanticipated during design implementation and verification of the design. The alternatives are to respin the part, which with today's mask costs is an additional half million dollars, or to reconfigure the part at bring-up. The latter is far less expensive. Unfortunately, it is impossible to anticipate where the bugs might be found and what can be done to fix them, but having more configuration options increases the likelihood that the next system-level bug can be fixed with reconfiguration rather than redesign.

Reconfigurable Logic

When deciding whether to use reconfigurable logic in a derivative design, three conditions of use are:

- If a design needs a number of different but only occasionally swapped applications, each one of which can fit in the FPGA block provided, and the performance of each of them requires that they be either in hardware or FPGA. This is because FPGA logic is 20 to 50 times larger than standard cell design. If only two or three well-defined applications need the FPGA, they can all be created in hardware and multiplexor-selected between them at far less silicon area than the FPGA logic requires.
- If designs have a number of unknown applications that can fit into the area and need the performance of an FPGA. In this case, the instant reconfigurable FPGA would be better, since size is less critical with that option.
- If the design needs a rapid prototype that can fit in a single chip. Some hand-held devices might fit that requirement. In this case, the FPGA logic is added to the design instead of the peripheral VCs. This is then used in the same manner as the rapid prototyping approach. The hardware kernel should be general to increase the likelihood of reuse, because this approach is much more expensive given that the FPGA-based design needs to be first debugged to use it to debug the derivative design.

Selecting Verification Methods

The belief that it is best to test thoroughly at each level because fixing a bug at the next is much costlier might no longer be true, because the cost of testing could outweigh the advantage of finding the bug. In semiconductor manufacturing, if the part is cheap enough and the yield is high enough, no wafer sort is done. The parts are just packaged and then tested. If they are bad, the package cost is lost. If the yield is 95 percent, the package can cost almost 20 times the cost of the test, and it is still cheaper not to do the test.

In platform integration most of the design is reused. Only the top level interconnect is new. As a result, most of the effort is in verification. The simulation speeds for a design go from architectural, which is the fastest, to physical netlist, which is the slowest. Emulation and rapid prototyping are as fast or faster than architectural simulation. If the team is constant over the life of the project (possibly true for platform integration, not design), every day that the simulation can run faster saves a day of development at the project's run rate. For example, if a bug at RTL costs a day to find and fix, at the netlist level it costs ten days.

If the simulation necessary to find the bug costs over ten days to run, it is cheaper to let the bug get caught at the later level. Follow these guidelines when determining which type of verification to use and when:

- Do as much simulation as possible at the architectural level, because that is when bugs are the cheapest to find and correct, and the simulation is efficient.
- Do successively less simulation as the design progresses down to physical netlist. Never do more than it takes to fix the bugs at the next level. It is safe to simulate less than five times as many hours as it took to fix the bugs at each level, because each level is expected to have fewer bugs than the previous level (about half), and leaving those bugs is acceptable if simulating at this level costs more than catching them later.
- Do as much rapid prototyping as reasonable, but no less than the amount of architectural-level simulation, because rapid prototyping is faster than architectural-level simulation, and the cost of bugs after fabrication is more than ten times the previous level.
- Drop simulation up to the last level if bugs are not found at the previous level, because simulation is slower at each successive level. The same amount of simulation time checks less at each successive level. If the current tests do not catch bugs, the most likely outcome is that bugs will not be found at the next level. Rather than increase the test time, skip to rapid prototyping to verify the design.

These guidelines do not apply to any new logic introduced at this level. For example, if the VC interface logic is added at RTL, the logic for the VC interface must be tested. If bugs are not found, it does not need to be tested until rapid prototyping. If rapid prototyping finds a bug, you must still go back to the appropriate level and simulate to isolate the bug. This approach, as radical as it sounds, will probably save time in most derivative design developments.

Moving Forward

The techniques and requirements for defining, implementing, and verifying a derivative design are evolving as the technology grows and the markets shift. Today, derivatives can also be created within narrow application spaces by mixing predesigned, parameterized interfaces with state of the art hardware and software EDA tools. This section discusses ways to evolve a current reference design into a full platform, as well as a number of design strategies.

Building a Platform Library

Realistically, an integration platform is not created until a design suggests that a platform is needed. So what should the hardware kernels have in this platform? Since only one design exists to base any real analysis on, the platform library should contain at least one very general hardware kernel. Initially, the best structure is a flexible hardware kernel, with all the other VCs available as peripheral blocks. Because it is very expensive to qualify a hardware kernel and all the VCs in a platform library, they must be designed for reuse. The most flexible hardware kernel will be reused if there are any more derivative designs. After a number of derivative designs have been successfully completed, the common elements of all the derivative designs can be incorporated into a new hardware kernel to reduce the need to continually integrate these elements on successive derivative designs.

If a derivative design requires a new function, the function should be implemented as a peripheral VC and added to the platform library prior to designing the derivative. As new peripheral VCs are added, more derivative designs can use them, which creates a demand for a hardware kernel with that function built in. This is one way the platform shifts to better covering the market segment it is addressing.

The time to build another hardware kernel is when the estimated savings from having the new hardware kernel is greater than the cost of developing and qualifying it. As the number of derivative designs that have a common set of functions increases, so does the number of test suites and models that can be used to create a hardware kernel, so the cost of developing the hardware kernel should come down over time.

A derivative design can be viewed as the smallest combination of otherwise common functions in other derivatives. Converting an entire derivative design into a hardware kernel is less costly than creating one from the common functions found in a number of derivatives. In this case, additional VC interfaces should be added to the newly created hardware kernel. Over time, and with continual increases in semiconductor fabrication, derivative designs will become hardware kernels, thus creating new derivative designs again. This trend keeps the integration level of the VCs in the platform naturally growing with the capabilities to fabricate the designs.

Migrating Software to Hardware

Throughout this book, we have described general strategies and methodologies for making platform-based designs. Although these methods are broadly applicable, many of the tools and methods needed to address the strategies discussed are either embryonic or non-existent. However, some highly focused methodologies, using the procedures described, can be created today for developing derivatives for specific market segments. One such method is migrating software to hardware.

The key for making platform-based systems is to provide predesigned interfaces for custom peripherals. The assumption is that these peripherals start out as software blocks, which are then transferred into hardware with a number of interface options, such as separate memory addressed/interrupt controlled peripherals, co-processor blocks, and execution units for special instructions.

The starting point for this could be an application that runs on a prototype of the platform, with no custom logic. The entire design is in software, and the prototype is created from the hardware kernel. Special profiling tools are created for analyzing which software modules are candidates for conversion to hardware. Manual procedures must be created to define a block-based coding style for the software blocks. With these tools and procedures, the engineer can estimate the performance of the selected blocks. The specific type of interface for each of the blocks must be defined during this assignment process. Once the specific code and the type of interface is defined, behavioral synthesis tools are used to convert the software module into RTL code. After this, the logic is modified to include the protocols for the appropriate interface for each of the RTL blocks and to set the corresponding parameters on the parameterized hardware kernel. The design can then be mapped to the FPGA portions of the rapid prototype or integrated into the ASIC chip, which contains the hard blocks of the non-parameterized portions of the platform.

After the design is sufficiently verified on the rapid prototype, it runs through a simplified RTL to silicon flow. The final simulation and timing are done on a back-annotated, gate-level design of the custom logic, mixed with a timing-accurate model of the fixed portions of the platform.

The beauty of this system is to let the behavioral synthesis tool do what it is best at: convert the software into hardware. On the other hand, the VC interface, interrupts, and hooks for the co-processor interface are hardware headers that get added to the synthesized design, along with generators for the drivers. In all, the process replaces a software module with a hardware call that invokes the same function in hardware. By designing a specific set of tools, hardware protocol interfaces, and software interfaces, a system can be created using tools that are available today when a more general approach is not possible.

Similar systems could also be built to convert hardware to software on a module by module basis.

Adaptive Design Optimization Strategies

Many books have been written about the adaptive capabilities of neural networks[1] and genetic algorithms[2] in recent years. As work progresses in these areas, it will be possible to apply these techniques not only to aid the design develop-

---

1. Yoh-Han Pao, *Adaptive Pattern Recognition and Neural Networks,* Addison Wesley, 1989.
2. David E. Goldberg, *Genetic Algorithms,* Addison Wesley, 1989.

ment process, but also in optimizing systems while they are in operation. One example of this, which was introduced in Chapter 5, is in the situation where the bus architecture includes a programmable arbiter: a separate monitor could keep track of the system operations and dynamically adjust the operation of the arbiter to improve the operation of the bus.

Other cases could be constructed when using a system with embedded FPGA logic. For example, when using a chip that has both a processor and reconfigurable FPGA logic, a specific application could redundantly consist of both FPGA hardware and software blocks that do the same task. It is difficult to schedule the hardware tasks efficiently. It may take many cycles to load a hardware task into the FPGA's configuration memory. If the FPGA logic has a configuration cache with many planes of configuration memory, there is also the question of when to load the memory. A set of decision trees could quickly lead to more alternatives than can be stored within the configuration cache. In these situations, adaptive algorithms could decide which hardware and software modules to use and when to load them. With this approach, it is easy to imagine a system that can tune itself to meet the performance requirements if given enough learning time.

Taking this approach one step further, it might be possible to apply "genetic trials" on multiple copies of such devices. Selecting the best results to reproduce on the next cycle of trials would lead to the correct scheduling approach. The resulting hardware and software modules could be frozen into the reduced set of software and hardware modules actually used, or they could be converted into non-FPGA hardware and software for an even smaller design.

Mixed Electronic-Mechanical Designs

With the rapid progress being made in microelectronic mechanical systems (MEMS), it is not hard to imagine using these devices on SOCs. The economics of integrating MEMS is similar to other process-specific devices, such as analog, DRAM, or EEPROM. If the MEMS takes up a large part of the chip, integration might be economical. Otherwise, the process variation must be small, since the process cost is applied to the entire chip. Fortunately, as the standard CMOS process becomes more complex, creative micromechanical engineers are apt to find ways to create mechanical structures using the same process steps with only minor variations.

In any event, these blocks can be integrated using the same procedures as those used for integrating analog devices. In fact, for most sensors, an analog component is necessary to sense or create the electrical signals generated from or sent to the mechanical devices. In other words, the mechanical devices should have analog control and sense logic with digital interfaces to connect with the rest of the SOC in the same manner as the analog blocks.

As more MEMS are created, entire platforms will be designed around certain key mechanical devices. For example, acoustic devices can now be integrated on the chips, enabling hearing aids to be implanted directly in the ear. Optical mirrors have been created to view overhead projection of computer output. One of the easiest mechanical devices to create is accelerometers, which are being used to interpolate between location measurements in global positioning systems (GPS). A future, single-chip GPS system could include a processor, high-speed analog receivers, and accelerometers.

8

Analog/Mixed-Signal in SOC Design

Many of the components along the periphery of SOCs will be the analog interfaces to the outside world. The pen and display interface on a personal digit assistant, the image sensors on a digital camera, the audio and video interface for a set-top box, and the radio frequency (RF) interface on a portable Internet device all require analog/mixed-signal (AMS) components. It is projected that out of all design starts in 1999, 45 percent will contain AMS design content.[1] The percentage of design starts in 1997 with AMS was 33 percent.

This chapter presents the major issues surrounding AMS in SOC, and illustrates a methodology for AMS block authoring, block delivery, and block integration.

In terms of the tasks associated with the platform–based design (PBD) methodology, this chapter discusses the tasks and areas shaded in Figure 8.1. Not surprisingly, these are similar to the digital hardware tasks. However, some tasks, such as static verification, have less meaning in AMS block design and, therefore, are not shaded.

Using AMS Components in SOCs

Integrating AMS components poses significant design challenges and added risks. The challenges mainly lie in the fact that unlike a digital signal where the information is encoded as either a 0 or 1 (a low voltage or high voltage), the information contained within an analog signal is continuous to an arbitrary degree of resolution depending on the component. Therefore, an AMS component and all of its interfaces are far less immune to noise in the system

---

1. Dataquest survey of design starts, actual and planned, September 1997.

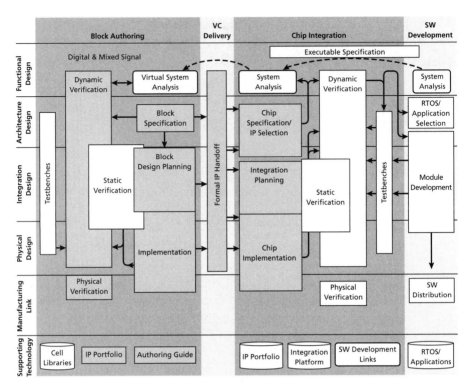

Figure 8.1. PBD Methodology: AMS in SOC Tasks

compared to their digital counterparts. The main hurdle, therefore, in integrating AMS components is to account for their more sensitive nature.

Creating AMS components also presents significant design challenges. AMS blocks are consistently faced with issues, such as precision, noise sensitivity, high frequency operation, high accuracy operation, and the need for a high dynamic range within increasingly lower power supply voltages. Design hurdles are transient noise, substrate coupling noise, ground resistance, inductance, and other effects. Many techniques have been developed through the course of integrated circuit (IC) design history to overcome these hurdles, but the challenge now is to collect and combine these techniques effectively in a new context, the SOC.

Some of the main issues for AMS in SOC addressed in this chapter are:

• What type of AMS components are most appropriate for SOCs?
• How do I decide whether an AMS component should be on-chip or off-chip?
• How do I design AMS virtual components (VC) so that they work in SOCs?
• How do I author AMS VCs so that they are reusable?
• How do I hand off an AMS design to a digital system integrator who does not understand AMS issues?

- How do I successfully buy AMS VCs?
- How do I perform true system design of AMS blocks making trade-offs at all levels?
- How do I integrate (verify) AMS VCs successfully?
- What modifications need to be made to a VC portfolio to account for an AMS VC?
- How do I successfully integrate AMS blocks into a platform-based design?

AMS Defined

Depending on the context, the terms analog and mixed-signal often have slightly different meanings. In this book, analog refers to designs that contain signals, either voltages or currents, that predominately have continuous values. Examples of analog design would be continuous time filters, operational amplifiers, mixers, etc. Mixed-signal refers to designs that contain both analog and digital signals, but where the design is mainly focused on the analog functionality. Examples of mixed-signal design are analog to digital (A/D), digital to analog (D/A), and phase-locked loops (PLL). AMS refers to the combination of analog and mixed-signal designs. A typical SOC would be described as digital with some AMS components because, although such an SOC does contain both analog and digital signals, the focus of the design and verification for the system is digital. The AMS focus is primarily at the block level only.

In the simulation context, mixed-signal is defined as mixed-circuit and event-driven simulation. The AMS in Verilog-AMS or VHDL-AMS indicates that the language allows the user to model some parts for a circuit simulator and other parts for an event-driven simulator. In the manufacturing test context, a typical SOC with AMS components is referred to as mixed-signal because in looking at the part from the outside, both the digital and AMS design portions require equal attention. The analog components must be tested to see whether they adhere to parametric values. The digital components require static and dynamic structural testing. Sometimes, the IC looks as if it were completely analog, because the signals are almost exclusively analog-in and analog-out. The digital circuitry is only used for intermediate operations within the chip.

What Is an AMS VC?

AMS VCs are intellectual property (IP) blocks, such as A/D and D/A converters that have been socketized. Common confusion often lies in that many consider VCs to imply reuse in the digital sense. AMS VCs are not reusable in the same way soft or firm digital VC cores are reusable. Synthesis does not exist for AMS blocks. Unlike Verilog or VHDL, which have a synthesizable subset, the emergence of AMS languages only allow modeling the behavior of AMS blocks, not their synthesis. Today, AMS VCs are delivered solely in hard or layout form. They do have value and can be designed to be reusable (discussed later in this chapter).

AMS in SOC

AMS in SOC, as used in this book, refers to putting AMS VCs in typical SOCs. Figure 8.2 shows an example of AMS in SOC. Component-wise, the design is predominately digital with embedded software running on a microprocessor. It has a system bus and peripheral bus. The video interface, the audio coder-decoder (codec), the PLL, and the Ethernet interface are AMS VCs. It is also an example of an SOC that could be analog-in, analog-out. A video stream could enter via the 10Base-T interface, be decoded via the digital logic and software, and sent to the video interface for output. This can be contrasted to what others have also called mixed-signal SOCs, which are referred to in this book as AMS custom ICs.

Table 8.1 illustrates the key design differences between AMS in SOC and AMS custom ICs. There are, of course, many other domains of AMS design, such as high frequency microwave, power electronics, etc., which are not discussed in this book.

The AMS VCs in SOC are generally lower in performance than the AMS blocks found in custom ICs. Two solutions are available to create higher-performance blocks in SOCs. For medium-performance blocks, the AMS VCs are integrated into the hardware kernel, or for higher performance, the analog functionality is left off-chip. If the AMS design can use either solution, one risk

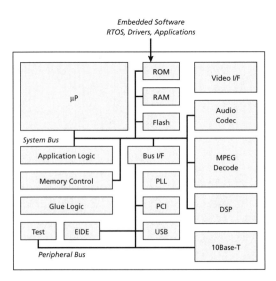

Figure 8.2. Example of AMS in SOC

Table 8.1. AMS in SOC vs. Custom ICs

Design Characteristics	AMS in SOC	AMS Custom ICs
Typical IC differentiator	Complete set of features	Analog performance
Analog on IC?	Only if benefits overall system	Yes, to meet function and performance objectives
Technology	CMOS (< 0.35µm) for density reasons	Any
Example designs	AMS block: Ethernet physical layer, A/D, D/A, PLL Chip: Set-top box	PRML (Partial-Response Maximum-Likelihood), xDSL (Digital Subscriber Loop) front end, RF
Separate authoring and integration?	Yes, blocks preverified with simple defined I/O	No
Primary design challenge	Interfacing within system and design for robustness in noisy environment	AMS performance
Design reuse	VC, source- or core-based hard blocks	Knowledge-based
Designers	AMS block: AMS designers Chip: system designers	AMS designers
AMS design focus	Commodity-type blocks (A/Ds, D/As, PLLs) where context is not well known	Leading edge performance; full exploitation of technology
View of system	Digital system	Analog
Verification level	Mixed hardware/software, analog/digital only as necessary	AMS simulation, bottom-up verification with behavioral models
Usefulness of VSI or equivalent	Necessary	Not necessary
Starting point in building design methodology	ASIC digital design methods	Analog design methods

reduction technique is to design one copy on-chip, but leave pins available to have a second copy as an off-chip component. If it fails on-chip, the product can still be shipped. The fully integrated solution can be deferred to the derivative design, if necessary. The decision is ultimately driven by cost and time to market (TTM) considerations.

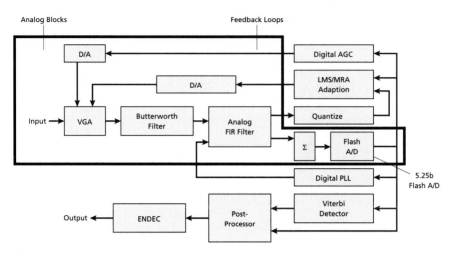

Figure 8.3. Example of AMS Custom IC

AMS Customs ICs

Figure 8.3 shows an example of an AMS custom IC, a CMOS Extended Partial-Response Maximum-Likelihood (EPRML).[2] It is characterized by many feedback loops between the blocks at the first level of decomposition of the IC. The design of these blocks is very tightly coupled to the process, to each other, and to the system in terms of function and I/O. The analog or mixed-signal blocks also make up a significant percentage of the number of blocks on the chip.

Over time, as process technology improves, many of the functions implemented by the AMS custom ICs will become VCs in SOCs. A 16-bit audio codec was once a standalone part. Today, it is being integrated into SOCs. Functions that tend to follow this transition are those where performance specifications (for example, frequency of operation, dynamic range, and noise) are fixed. An RF front end is a good example of a function that will not follow this transition. Although low-performance RF blocks have been designed into SOCs, market forces continually push the frequency of operation of RF front ends higher and higher. RF front ends, therefore, tend to remain as separate ICs requiring special processing steps and methodologies for custom design.

2. J. Fields, P. Aziz, J. Bailey, F. Barber, J. Bamer, H. Burger, R. Foster, M. Heimann, P. Kempsey, L. Mantz, A. Mastrocola, R. Peruzzi, T. Peterson, J. Raisinghani, R. Rauschmayer, M. Saniski, N. Sayiner, P. Setty, S. Tedja, K. Threadgill, K. Fitzpatrick, and K. Fisher, "A 200Mb/s CMOS EPRML Channel with Integrated Servo Demodulator for Magnetic Hard Disks," *ISSCC Digest of Technical Papers*, SA19.1, 1997.

Block Authoring

The role of AMS block authoring is to design blocks that meet the performance specifications within the SOC environment in an efficient and low risk manner. Unlike digital block design, where automated register transfer level (RTL) synthesis flows exists, AMS design is accomplished by manually intensive methods. The majority of design automation tools available are focused on design capture and verification. Therefore, it is in the design methods that VC providers differentiate themselves in terms of rapid design, reuse, and repeatability.

To design within performance specifications in an SOC environment, the VC provider must design for robustness and compatibility within the SOC process technology. The design must work in a noisy digital environment and not use analog-specific process steps, such as capacitors or resistors, unless allowed. Usually this is accomplished by not pushing performance specifications. For fast retargeting, VC providers often write their own design automation tools, such as module generators for specific circuits. However, much of the success is from designing in a more systematic fashion so that each design step is captured in a rigorous way. Designing with reuse in mind is also required. Often, designs are easily retargeted because a designer, knowing that the design would be ported to other processes, did not use special process-dependent techniques to achieve performance.

As shown in Figure 8.4, the AMS design process begins with specifications that are mapped to a behavioral model, where parameters are chosen for the basic building blocks. These parameters become constraints to schematic block design. The schematic is then mapped to layout or physical design. Some amount of automation can be applied, or scripts can be written to allow for fast reimplementation of certain steps, but in general the design process is a custom effort. Verification occurs at all levels to insure that the design specifications are met. Once fully constraint-driven and systematic, this serves as an ideal methodology for AMS VC design.

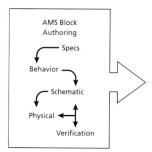

Figure 8.4. AMS Design Process

Authoring AMS VCs for Reusability

To reuse AMS VCs, a systematic, top–down approach, starting from the behavioral level and based on early verification and constraint propagation, needs to be employed. Currently, AMS block design faces the following challenges and issues:

- Chip-level simulation requires too much time.
- Design budgets are not distributed in a well-defined manner across blocks.
- Too much time is spent on low-level iterations.
- Design is not completely systematic.
- There is limited or no use of hardware description languages (HDL).

The top-down, constraint-driven methodology addresses these issues. Figure 8.5 shows an overview of this methodology.[3] Specifications for the block to be

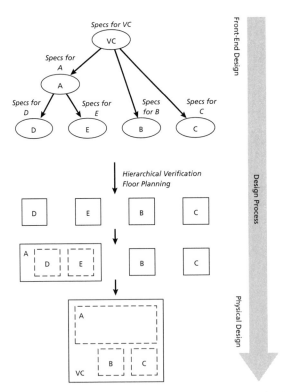

Figure 8.5. Top-Down, Constraint-Driven Design Methodology

3. For a full description of this methodology, refer to H. Chang, E. Charbon, U. Choudhury, A. Demir, E. Felt, E. Liu, E. Malavasi, A. Sangiovanni-Vincentelli, and I. Vassiliou, *A Top-Down, Constraint-Driven Design Methodology for Analog Integrated Circuits*, Kluwer Academic Publishers, 1997.

designed enter from the top. An appropriate model, either in a HDL or schematic, is built for that block where the specifications for the next level are the parameters or variables in that model. The goal of the model is to translate these parameters into its specifications, so that various decompositions can be tried to find an optimal decomposition that meets the block requirements.

The following equation is used for mapping from one level to the next:

$$\max flex(var)$$
$$\text{s.t. } perf(var) + \Delta specs <= specs$$

At each design level, flexibility (*flex*) of the next design step is maximized subject to the specifications for that block. The variables (*var*) are the performance specifications (*specs*) for the next level in the hierarchy.

This mathematical representation captures the design intent in a systematic method, enabling design reuse. It also provides some degree of automation to decrease design times. Using a mathematical program solver increases the speed of the mapping. The mapping process continues until the full schematic design is completed. A similar method is used for layout design.

This methodology and the function-architecture co-design approach described in Chapter 4 are based on the same goals and execution. As in the function-architecture co-design approach, the top-down, constraint-driven design methodology has the following characteristics:

- Captures the design process and makes it more systematic.
- Has rigorous definitions behind each step, and creates intermediate stopping points or sign-offs to enable different final implementations to be retargeted.
- Begins with a model that is independent of the final implementation so as to target different design contexts and applications.
- Provides the methodological context for decisions to be made early in the design process, unencumbered by unnecessary details of implementation.
- Compares system behavior in the abstract models to map the behaviors to architectures.
- Allows for reuse of pre-existing components that fix parameters in the behavioral representation.

Support tools for this method include mixed-level, mixed-mode simulation capabilities; constraint-driven, semi-automated to automated mixed-signal layout place and route at both the block and device level; budgeting/optimization tools; verification and hierarchical parasitic extraction capabilities; design data management; design automation software libraries to support building relatively simple module generators for fixed-structure components; and statistical simulation packages to aid both behavioral simulation as well as design for test.

Fundamental tools include tools for substrate coupling analysis because of the noisy digital environment; high-level language capability; schematic capture; layout capabilities; and verification tools, such as design rule checkers and corner checking.

Using Firm AMS VCs

Although a synthesis methodology does not exist, a systematic design method is sufficient to discuss VCs that are more retargetable than hard VCs. Figure 8.6 illustrates a higher-level handoff. If both the VC provider and VC integrator have an understanding of the same systematic design methodology, the VC provider can pass the intermediate design data to the VC integrator to finish the layout.

However, it is more likely that the VC provider keeps the firm information and takes advantage of the format for fast retargeting to different manufacturing processes to deliver hard VCs. The provider can even build this firm VC into a VC portfolio or integration platform for more reuse.

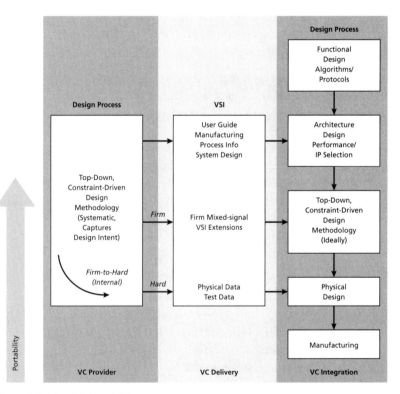

Figure 8.6. Firm VC Handoff

Past Examples

Firm-like IP for AMS is not unprecedented. The term IP is used in this context to indicate that the block being discussed has not been socketized. Much of it has been in the form of module generators, which include complete design-to-layout generators[4] to layout-only generators.[5] Knowledge-capture systems that attempt to encapsulate IP have also been created.[6] In general, these pieces of IP consist of fixed architectures, allowances for minor performance specification changes, allowances for process migration, detailed layout information, and automatic synthesis from specifications.

These module generators and knowledge-capture systems have had only limited success because of two fundamental flaws. First, there is no standard for the final set of deliverables. The IP integrator does not know what information is in the IP, because of a lack of definition starting from I/Os and operating parameters all the way to the configuration management of the IP creation tool. Secondly, and a more fundamental flaw that is not so easily addressed, is that the firm IP has similar characteristics to silicon compilers (see Figure 8.7), and thereby, suffers some of the same problems, such as generators that are

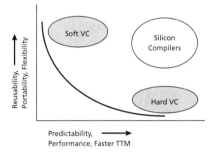

Figure 8.7. Comparing Silicon Compilers to Soft and Hard VCs

4. H. Koh, C. Sequin, and P. Gray, "OPASYN: A Compiler for CMOS Operational Amplifiers," *IEEE Transactions on CAD*, February 1990; G. Jusuf, P.R. Gray, and A.L. Sangiovanni-Vincentelli, "CADICS-Cyclic Analog-to-Digital Converter Synthesis," *Proceedings of the IEEE International Conference on Computer-Aided Design*, November 1990, pp. 286-289; and R. Neff, P. Gray, and A.L. Sangiovanni-Vincentelli, "A Module Generator for High Speed CMOS Current Output Digital/Analog Converters," *Proceedings of the IEEE Custom Integrated Circuits Conference,* May 1995, pp. 481-484.

5. H. Yaghutiel, S. Shen, P. Gray, and A. Sangiovanni-Vincentelli, "Automatic Layout of Switched-Capacitor Filters for Custom Applications," *Proceedings of the International Solid-State Circuits Conference (ISSCC),* February 1988, pp. 170-171.

6. M. Degrauwe, et al., "IDAC: An Interactive Design Tool for Analog CMOS Circuits," *IEEE Journal of Solid-State Circuits*, vol. SC-22, n. 6, pp. 1106-1116, December 1987; and J. Rijmenants, J.B. Litsios, T.R. Schwarz, and M.G.R. Degrauwe, "ILAC: An Automated Layout Tool for Analog CMOS Circuits," *IEEE Journal of Solid-State Circuits*, vol. SC-24, n. 2., pp. 417-425, April 1989.

extremely difficult to maintain. This results in fast obsolescence of the IP, and because it is extremely optimized for performance, it is difficult to re-tune for survival even for a few manufacturing process generations.

Instead, we propose a firm VC based on the top-down design methodology with a strict set of outputs, repositioning its use as shown in Figure 8.8. Firm AMS VCs contain a fixed architecture, layout information, and sufficient information to connect the architecture to implementation. This can be done in the form of constraints.

The AMS VC has a relatively fixed function. For example, an A/D converter remains an A/D, but the number of bits can vary. Performance specifications can also vary within the scope of the architecture. They can include portability of technology (manufacturing), operating conditions (supply voltages, currents, etc.), external layout-considerations (VC I/O pin locations, etc.), and optimization target (area, speed, power, cost, etc.).

When using firm VCs, the user can be either a VC integrator or a VC provider who delivers hard VCs, but must be familiar with the top-down, constraint-driven design methodology. The user must also be competent in AMS IC, block, or VC design. Because the top-down design methodology is not an automatic synthesis process, it is not anticipated that the VC user will have an automated design process once the handoff has been made. To bridge the differences in knowledge and design processes between VC providers and VC integrators, application notes and applications engineers need to be provided.

There is an art to AMS design and verification, since there is no automated process for doing this. For example, part of the handoff might specify that a component, such as an operational amplifier, must be designed to a certain set of specifications. It is the VC user's job to design and verify this block to specifications. As another example, a question could arise as to how much verification is sufficient. Is parasitic layout extraction plus circuit simulation sufficient? How are offset variations verified? Although the constraint-driven layout tools remove some of the art in answering these concerns, there are still holes that require an AMS designer to catch.

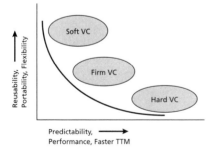

Figure 8.8. Soft, Firm, and Hard VC Trade-offs

Depending on how critical a block is, varying levels of specifications can be used. Unlike digital firm, which is strictly based on gate-level libraries, the analog design hierarchy does not stop until the device (transistor, resistor, capacitor) level. Thus, if a block is critical, there may be a descent for that block to the device level. When the descent is not to this depth, the remainder of the design is treated as a black box. The methodology does also allow for stopping points when standard analog cells are available.

Basically, the key to this is to save an intermediate state of the design. To do so, the hierarchy must be described as shown in Figure 8.5 and each block must be labeled. All of the components used in the mathematical equation also need to be described. Having the right behavioral models is critical to the design methodology. In general, ranges for the specifications should be given, so that a sense of reasonable constraint values are available. An initial value for the $\Delta spec$ budget should also be provided. Finally, application notes should be given on methods for solving the mathematical program. Often specific optimization algorithms must be tuned or entirely different algorithms need to be applied to solve a particular problem.

In terms of physical design, additional specifications have to be included. These are specifications for how to build the VC, not how to integrate it, and include standard placement, routing, and compaction constraints. They can be very detailed, or they can be left to the designer or tool to derive based on the constraints and allowances for performance degradation ($\Delta spec$).

AMS VC Delivery

How the block is delivered to the VC integrator after it has been designed is critical to the successful use of the VC. To simplify this discussion, only the delivery of hard VCs is discussed here. This section focuses on the technical aspects of what is required for delivery as outlined by the Virtual Socket Interface (VSI) Alliance. Several key operations and business factors that must be considered are also touched upon.

VSI Alliance's AMS Specifications

The VSI Alliance's Mixed-Signal Development Working Group (MS DWG) presents in its 1998 work, *Analog/Mixed-Signal VSI Extension* (AMS VSI Extension), standards for specifying the technical requirements for VC delivery. The MS DWG extends the digital VSI specifications to account for the added design challenges presented in AMS design.

The methodology context for the AMS VSI Extension, as shown in Figure 8.9, looks only at the delivery of hard VCs, which is far simpler than what is required for soft or firm VCs. Because the VC integrator does not have to design any of the VCs, the intermediate information that would be required for

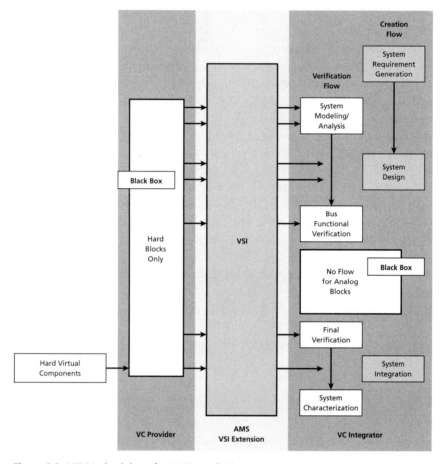

Figure 8.9. VSI Methodology for AMS Hard VCs

soft and firm does not enter into the picture. The only places for information exchange are at the top of the flow for system architecture and system design information and at the bottom of the flow for the actual physical layout information. This also enables VC providers to use any methodology they want for design. The AMS VSI Extension does not dictate the methodology employed by the AMS designer; it only specifies the information that needs to be transferred.

Because of the similarities between a digital hard VC and an AMS hard VC, the AMS VSI Extension follows the definition for hard VCs for deliverables where there is a digital counterpart.[7] Examples from the Physical Block Implementation Section are shown in Figure 8.10. This section contains all the nec-

---

7. "Structural Netlist and Hard VC Physical Data Types," VSI Alliance *Implementation / Verification Development Working Group Specification 1,* Version 1.0, June 1998.

Section	Deliverable	VSI Format	AMS Hard
2.6	Physical Block Implementation		
2.6.1	Detailed Physical Block Description	GDSII	M
2.6.2	Pin List/Pin Placement	VC LEF	M
2.6.3	Routing Obstructions	VC LEF	M
2.6.4	Footprint	VC LEF	M
2.6.5	Power/Ground	VC LEF	M
2.6.7	Physical Netlist	VC Hspice	CM

Figure 8.10. Example of VSI Hard VC Deliverables

essary information regarding layout. The first column refers to the item number, and the second contains the name of the deliverable. The third column specifies the selected format in which the deliverable has to be transferred from VC provider to VC integrator. The fourth column shows whether the deliverable is mandatory (M), recommended (R), conditionally mandatory (CM), or conditionally recommended (CR). The conditional clause is described in detail in the VSI specifications. For example, deliverable 2.6.1, "detailed physical block description," is the layout itself, and it is mandatory that the VC provider delivers this to the VC integrator in the GDSII physical description format.

For hard VCs, the digital blocks are not concerned with physical isolation techniques such as substrate well-rings. However, these are often used in AMS blocks. Thus, in the AMS VSI Extension, Section 2.6.A8.1 has been added (see Figure 8.11) so that the AMS VC provider can give information to the VC integrator about what is necessary for physical isolation.

The VSI specification can also be used in other ways. It is likely that it contains a superset of what most design groups are already doing when exchanging VCs. In this case, a current methodology can be replaced by the VSI. A VC integrator can use the VSI to prepare the internal design process to accept VCs

Section	Deliverable	VSI Format	AMS Hard
2.6.1.A1	Layer Mapping Table	document	M
2.6.A8.1	Physical Isolation Specifications	document	M
2.6.A9.1	Placement Constraints	document	M
2.6.A9.2	Proximity Effects	document	M
2.6.A10.1	Special Hookup Guidelines	document	M
2.6.A10.2	Routing Constraints	document	M
2.6.A10.3	Special Pin Requirements	document	M
2.6.A10.4	Additional power, ground, and substrate interconnect constraints	document	M

Figure 8.11. Example of AMS VSI Extensions

that would be delivered in the VSI format. Both VC author and integrator work in parallel to produce and accept VSI parts that can be quickly integrated.

Operations and Business Issues

A key operational issue in VC delivery is the means of delivery and the applications-engineering support that goes with it. The VC integrator is typically not an AMS designer. Any modification to the VC requires significant assistance from the VC provider, since even small changes can have a large impact on the block in terms of function and overall AMS performance specifications. Small modifications also usually result in re-running long verification cycles. Often, in selecting VCs from a variety of vendors, a suitable service model is a prime requirement. Some business issues to consider when selecting VCs include:

- *Patents* Most of the popular standard circuit topologies are patented. Small VC providers tend to have little in the way of a patent portfolio, and rely on the buyer of VCs for patent protection.

- *Pricing models* The VC provider and VC integrator must work out a pricing model for the VCs. Different models include a full license for VCs so that the buyer can use it anywhere anytime, a charge for the nonrecurring engineering (NRE) costs, and payment based on royalties. Typically, a combination of these is used.

- *Protection* It is a good idea for the integrator to obtain as much information about the VC as possible to mitigate integration risks. Often it is demanded as an insurance policy to the design know-how. This must be balanced with the concern that it is prudent to release as little information as possible.

- *Cost* Projections show that the price of a SOC is not likely to increase even though the number of VCs in a SOC will increase. This will drive the necessity for low cost VCs. The VC provider must have a way to counter this.

- *Time for negotiations* This is often referred to as "time-to-lawyer." In some cases, settling a contract requires more time than the design.

- *Margins* For a VC provider, a key to growth and survival is to create high margin products and to not just deliver design services on a time and material basis. This is especially true for AMS design, which is very designer intensive. Without margin, VC providers cannot bootstrap themselves for growth. Often they look for other product lines to boost their overall margin, such as selling home-grown computer-aided design tools, which, if done correctly, can be of higher margin.

- *Vendor status* Being selected by a VC integrator often means that the VC provider becomes a qualified vendor, which is often far beyond the capabilities of the VC provider. As a vendor, some of the issues the VC provider needs to take into account include acquisition, integration, support, management, easy-of-use, leverage, and optimization.

- *Qualification* This might be difficult for VC integrators, since they might require their own AMS designers for qualification and certification of AMS VCs.

AMS Components in SOCs

The key to SOC VC selection, chip integration, and verification is for these processes to occur in as similar a manner as possible as when using digital VCs.

System-Level Design and VC Selection

In the area of system-level design that is software and digital hardware dominated, where automation tools and flows are almost non-existent for AMS components, and where AMS components are treated as mere functions existing on the periphery of the system, the most critical need in regards to AMS is understanding analog and digital implementation trade-offs in terms of power, performance, and risk. Pragmatically, this means not writing unrealistic specifications for the AMS components.

At the most basic level, it is important that a joint appreciation between the digital and the AMS designer exists. The typical digital designer believes that AMS is mysterious and wants nothing to do with integration. Ironically, when designers find themselves involved in integration, they tend to oversimplify the interface. On the other hand, AMS designers view digital as trivial to design, since it is a subset of AMS. Both design teams often try to solve and/or underestimate each other's problems, which makes it difficult to do system-level design.[8] The digital designer can integrate AMS components, but it does require some extra effort. The difference between digital in AMS and ASICs is that the digital logic is much more complex than in AMS custom ICs. It has been said that system-level design with AMS only requires an appreciation for AMS, whereas the design of those components requires mastery.[9]

---

8. "Introduction to the Design of Mixed-Signal Systems-on-a-Chip," part of the "Design of Complex Mixed-Signal Systems on a Chip" tutorial, *Design Automation Conference*, June 1998.

9. R. Rutenbar speaking in the panel "How Much Analog Does a Designer Need to Know for Successful Mixed-Signal Design?" at the *Design Automation Conference*, June 1998.

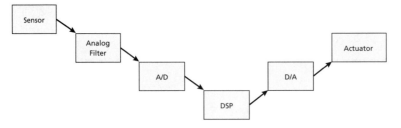

Figure 8.12. Electromechanical/Electronic Subsystem

Some of the most interesting issues in AMS SOC design lie in the trade-off analysis that determines the eventual integration level of an SOC device, and the particular choices which are made on how much of the design is implemented in different computational domains—electromechanical, traditional AMS, digital, or software.

Consider the electromechanical/electronic subsystem chain for an automotive application shown in Figure 8.12. If this subsystem is used for anti-knock engine control, it might consist of a sensor that converts the particular engine combustion conditions into streams of analog signals, which are then filtered for noise, converted to digital at a certain bit rate, and used as input for a control algorithm implemented in software on a digital signal processor (DSP). The control signals are converted from digital to analog and then used to control the actuators which directly impact engine operation.

Theoretically, all the components in this subsystem chain could be integrated into one device. For the components, it is possible to use either an expensive sensor, which produces a high-quality, low-noise analog signal, or a cheap sensor, which produces a low-quality noisy signal. Using a cheap sensor can be compensated by using either analog filtering or digital filtering in the DSP. Either of these would require a more expensive analog device (whether discrete or integrated), or a more powerful, more expensive DSP processor. Table 8.2 indicates possible trade-offs that would apply to either discrete or integrated SOC devices.

The solution that would be optimal for a particular application depends on the library portfolio of discrete devices (for an unintegrated option) or the

Table 8.2. Trade-off Possibilities

Configuration	Sensor	Analog Filtering	DSP
1	Expensive	Cheap	Cheap
2	Cheap	Expensive	Cheap
3	Cheap	Cheap	Expensive

available VCs for an SOC device that was going to integrate at least two stages of the subsystem.

When deciding the analog/digital partitioning and integration for a mixed-signal system, some options to consider include:[10]

- Custom mixed-signal ASIC or SOC
- Custom digital ASIC with external discrete analog devices
- Microcontroller with analog peripherals
- Standard processor or processor board and standard analog peripheral devices

Some of the criteria suggested by Small for making trade-off decisions include design risk (favoring standard parts), TTM (again, favoring standard parts, but also possibly to optimize to some extent with reusable VCs on an SOC device), performance (mixed-signal ASIC or SOC), packaging area (favoring ASICs over all standard parts), process issues (some analog functions do not work in digital CMOS processes), availability of analog VC functions commercially (this market is just emerging), test time (integrated SOC device versus discrete devices).

As with digital design, the trend toward highly integrated SOC devices incorporating analog VCs, mixed with significant digital devices, seems inexorable, albeit longer in coming than in the digital domain.

Making these decision requires AMS models. Early work by the MS DWG has postulated the need for three types of system models: system evaluation models, parameterized estimators, and algorithmic-level models. The system evaluation model represents the basic AMS function, such as A/D conversion or clock generation. At this stage, few attributes have been assigned to the function. A parameterized estimator is an executable model that characterizes what is possible for a particular function. For example, it can provide estimates in terms of power, frequency, and area as functions of performance specifications. The algorithmic level model assists in choosing specifications for the AMS functions. For example, it can be used to decide between a 6-bit and an 8-bit A/D converter. Note that these could be provided by a VC provider, or an SOC system designer might also have a set of these very basic models.

Chip Integration

Integrating the AMS block is relatively straightforward once the block authoring process and VC delivery process has been accomplished. In some sense, everything done so far has been to make this step as easy as possible. AMS blocks must be considered in terms of general clock, power, timing, bus, and test architectures, as well as general floor planning in terms of the block and the

---

10. Charles Small, "Improved topologies, tools make mixed-signal ASICs possible," *Computer Design*, May 1998, pp. 27–32.

associated AMS pads. Routing obstructions and constraints must be considered. Power and ground requires extra planning, since AMS VCs almost always require separate power pins, which are decoupled from the rest of the system. AMS blocks also contain additional placement, interconnect, and substrate sensitivity constraints.

In terms of chip assembly, it is just a matter of handling constraints such as those shown in Figure 8.11. Critical AMS nets can be implemented in the way of pre-routes, along with the other pre-routes for clock, bus, and digital-critical interconnects, power, and test.

Additional tool requirements include allowance for placement constraints in the floor-planning tool. For placement, an AMS block might be specified to be a certain distance away from the nearest digital block. The placement tool must also allow for this constraint. Geometric, electrical, symmetry, shielding, impedance control, and stubs might be specified for routing. The router must be able to support these structures. Additional power and ground wires can pose inconsistencies with verification tools. The overall tool system must be able to deal with these additions. Other requirements might be tools to handle the quiet power and ground rings for the pads that might be added.

The integration of digital blocks might also require using many of these techniques, especially in the routing. As process scaling continues, especially into the realm of ultra-deep submicron, fundamental characteristics of the process, such as gate delay and interconnect capacitance, do not uniformly scale to give benefits.[11] For example, gate delay tends to be inversely proportional to continued scaling, while interconnect capacitance tends to be proportional to scaling. In the context of SOC integration, critical factors affected by scaling include interconnect capacitance, coupling capacitance, chip-level RC delays as well as voltage IR drops, electromigration, and AC self-heat. Because of these factors, different scaling techniques are used to try to compensate for this degradation. However, in all of the methods presented by Kirkpatrick, interconnect capacitance, coupling capacitance, and global RC delays always degrade. The scaling method only changes the extent of the degradation.

Thus, for these SOCs, even if the IC only has digital VCs, methods for controlling these factors are critical. AMS constraint-driven design techniques can be applied to address these issues. The use of constraint-driven layout tools, as well as AMS simulation tools, can aid in automating place and route of VCs. For example, timing constraints using constraint-generation techniques can be translated to electrical constraints, which can finally be translated into geometric to perform the chip-level placement and routing.

11. D. Kirkpatrick, "The Implications of Deep Sub-Micron Technology on the Design of High Performance Digital VLSI Systems," Ph.D. Thesis, UC Berkeley, December 1997.

Verification Methods

Verification methods for hardware integration of AMS VCs focus on the interfaces to the AMS VCs. VC providers need to provide models to assist in this phase. Because a VC integrator follows an RTL-based design flow, AMS models need to be put into compatible representations so that the integrator can make it through the verification flows. These might not accurately represent the function or performance specifications of the AMS VC, but they do represent the necessary verification abstraction layers for VCs. An example would be an RTL-based digital placeholder or dummy model.

In terms of actual verification, the critical piece required of the AMS VC is the bus-functional model. This might need to be accompanied by a behavioral model representing the VC's digital behavior to be used in an overall chip-level verification strategy. Other verification models that might be required include a functional/timing digital simulation model for functional verification of the digital interface, a digital or static timing model for timing verification of the digital interface, and a peripheral interconnect model and device level interconnect model for timing.

AMS and Platform-Based Design

This section describes the role of AMS in platform-based design. Figure 8.13 shows how AMS fits in the transition to SOC. In synthesis-based timing-driven design (TDD), AMS does not play a role. In systems using TDD, analog functionality was added by having separate IC solutions. AMS blocks enter into block-based design (BBD), where blocks are integrated but require a lot of interaction between the block provider and block integrator. Finally, using AMS VCs in PBD has the advantage of a clean separation between VC author and VC integrator via a formal handoff.

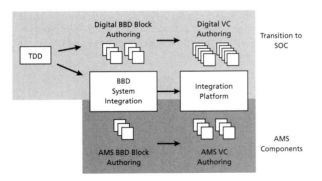

Figure 8.13. AMS and the Transition to SOC

With a few exceptions, using AMS VCs in PBD is similar to using digital VCs. One of the differences is that the VC database has to be extended to accommodate AMS VCs, and more information needs to be provided. Allowances need to be given for the performance specifications and the additional information required for using AMS VCs. In terms of the rapid-prototyping environment, AMS VCs need to be packaged as separate VCs and integrated at the board level. They cannot be mapped to field-programmable gate arrays (FPGA) as digital VCs can.

In addition, AMS can be included in the hardware kernel, which enables higher performance AMS blocks to be considered. Digital cores, which are typically non-differentiating for the AMS designer, provide additional functionality. The design techniques tend to be more custom and, therefore, can afford higher functionality. An example of this type of hardware kernel is a mid-performance RF front end. The digital blocks provide the interface functions required for the product, but the focus of the design is on the AMS portion.

In Summary

A key to success in using AMS in SOC is recognizing the design style required. This helps to ensure that the appropriate investments and techniques can be applied for the market in question. Using a methodology not suited for a design results in cost overruns, time delays, and design problems. If we return to the questions asked at the beginning of the chapter, we can propose the following solutions and approaches.

Table 8.3. AMS Solutions to Key SOC Issues

Issue	Recommended Solution
What type of AMS components are most appropriate for SOCs?	The components are typically limited to those that can be designed on an advanced digital, low-voltage process, and ones that are not so high in performance that they cannot be designed robustly in the noisy digital environment.
How do I decide whether an AMS component should be on-chip or off-chip?	If the block can be integrated to meet the TTM and cost constraints, it is put on chip. Otherwise, it is left off.

(Continued on next page.)

Table 8.3. (*Cont.*)

Issue	Recommended Solution
How do I design AMS VCs so that they work in SOCs?	The designer must trade off performance for robustness. The AMS VCs must have well-documented digital interfaces for communication to the digital environment. They must have proper considerations for integration and testability.
How do I author AMS VCs so that they are reusable?	A systematic top-down design approach can be used to capture design intent throughout the design process. This allows for more rapid retargeting of different design technologies. First and foremost, however, it must be designed with reuse in mind.
How do I hand off an AMS design to a digital system integrator who does not understand AMS issues?	In designing VCs, effort must be invested in a delivery methodology that includes using a VSI-like standard, which insures that all the information necessary for integration is provided. Application engineering support might also need to be provided to assist in integration.
How do I successfully buy AMS VCs?	A VC database can be used to find the VCs in question, but a variety of business issues must be addressed, including price, model, time for negotiations, cost, and VC qualification.
How do I perform true system design of AMS blocks making trade-offs at all levels?	Models can be provided by the VC provider to the system designer to allow for more accurate trade-offs in the function-architecture co-design phase of design.
How do I integrate (verify) AMS VCs successfully?	Models must also be provided by the VC provider at this level. These include static-timing models and functional RTL verification models.
What modifications need to be made to a VC portfolio to account for an AMS VC?	A VC portfolio must add analog performance specifications as well as the additional requirements as specified by the VSI Extension for AMS VC.
How do I successfully integrate AMS blocks into a platform-based design?	VCs must be prequalified prior to integration. All views must be checked, including bonded-out cores for the rapid prototyping system.

9

Software Design in SOCs

This chapter addresses the issues involved in software design and development for embedded SOC designs. In terms of the platform-based design (PBD) methodology introduced earlier, this chapter discusses the tasks and areas shaded in Figure 9.1.

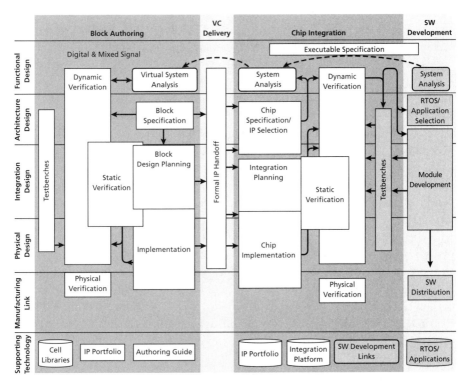

Figure 9.1. PBD Methodology: Software Design Tasks

Embedded Software Development Today

To develop a new approach to embedded software design, we should first look at what methodology is used today in order to examine its shortcomings. The availability of packaged development environments, such as board support packages (BSP), influences (or limits) which processor, real-time operating system (RTOS), and development methodology is chosen. Usually, the target RTOS tends to be well defined and often pre-chosen, possibly based on the previous development project in this family of products.

BSPs contain a target processor or processor core (packaged out), memory devices, a bus that is close to the target on-chip bus, slots for adding field programmable gate arrays (FPGA) that emulate hardware functions, and slots for adding other hardware functions encapsulated into IC form. They can also contain pre-packaged peripheral interfaces, and a connection to a debug and download environment running on a host workstation, very often a PC.

The RTOS can be downloaded onto the BSP with a variety of predefined configurations and configuration files. In fact, most commercial RTOSs have a host of configurable parameters so that they can be more precisely tuned to the specific application and to minimize latencies and memory consumption. Cross-compilers, host-based software development environments (debuggers, disassemblers, and so on), and host-based RTOS simulators enable code to be quickly developed, compiled, and debugged. In this sense, the BSP is very analogous to the rapid-prototyping environments for SOC integration platforms.

Cost (royalties for the core and RTOS; fabrication cost of the core), performance (of the processor and RTOS functions), and time to market (based on availability of the appropriate BSP and host-based development environment; reuse of existing software; and so on) affect which processor core, RTOS, and BSP are selected. These decisions are driven by the overall product design goals for function, performance, and development schedule. The product functions are defined and mapped to hardware or software via function-architecture co-design. Functions mapped to software are decomposed into components, which can either be new components or, preferably, existing ones. Often in parallel with this phase, existing reusable software virtual components (VC) from previous projects, a core vendor, or ones that can be purchased from third parties are chosen. Performance, cost, and the software's memory footprint are all important considerations. Sometimes existing components and libraries are chosen solely to avoid new development, even if they are more complex than required, a little slower than ideal, or occupy more memory than is targeted for the product application.

Existing reusable software VCs rarely cover all the product's required applications, so functions must be newly created. For that reason, it is extremely important to identify the software's architecture—the layering and dependencies of functions, and their interfaces—and to try to stabilize and freeze the interfaces as soon as possible to provide a stable base for new functions to be developed. These are then developed on the host, debugged (together with reused software functions), cross-compiled to the target processor, downloaded to the BSP, and integrated with other software VC components. The whole system is then available for prototyping, debugging, and optimization. The RTOS must be tuned to the application (for example, choosing stack sizes, detailed scheduling parameters, number of tasks if this is fixed by the application software, time slices, and any other configurable parameters) to minimize latency and memory consumption. Any specific device drivers for the intended SOC device must be written, optimized, integrated, and debugged in the system context.

In parallel with developing the new code and integrating it, the system testbench must be developed. The testbench can be used on the host-level debug and on the BSP during software integration and debug. It will also be used later with the manufactured SOC devices during final system integration on the real hardware.

During this process, if the system fails to meet performance requirements on the BSP, or exceeds planned integrated memory capacity on the SOC device, a well-defined procedure on how to correct the situation does not exist. Rewriting the performance-critical pieces of code to optimize the C or other high-level language code to minimize processor cycles and memory consumption is usually the first approach. Failing this, rewriting code in assembly level, or at least those parts of the code that consume the most processor cycles should be tried. If, as in many real-time embedded software applications based on digital signal processors (DSP), the code is already based on hand-coded assembly-level algorithms, there might be a fundamental mismatch between the DSP and the application requirements.

To identify the performance and memory-critical areas of the code, various kinds of performance-analysis capabilities need to exist in the software development environment, such as cycle-counting processor simulations (achieved by using an instruction set simulator (ISS) that is at least cycle-approximate); memory mapping and layout analysis tools; various kinds of breakpoints, flags, and triggers in the debugger; cross-probing of source code vs. object code; code profilers at the procedure, statement, and instruction level; and visualization tools for making sense of all the possible information that can be extracted.

The Architecture of Embedded Software

Typically, the architecture of embedded software is layered, as shown in Figure 9.2. Device drivers, which provide the basic hardware/software interfaces to specialized peripherals lying outside the processor, are closest to the

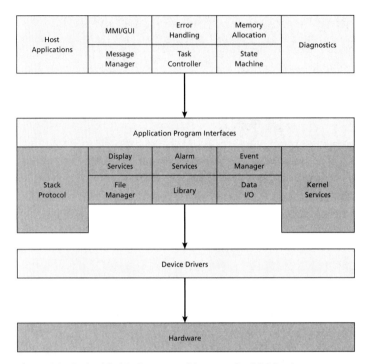

Figure 9.2. Embedded Software Architecture for SOC Design

hardware layer. Above that is the RTOS layer. It contains the RTOS kernel, which offers basic services such as task scheduling and intertask messaging. This layer also contains the communications stack protocol, which further layers communications between the application layer and the hardware layer, event and alarm trapping and management to handle hardware interrupts, and external IO stream management for files other data sources, displays, keyboards, and other devices. Higher-level system functions are made available to user applications and system diagnostics via Application Program Interfaces (API). The diagnostics and applications layer provide the highest level of control over the embedded software tasks, user interfaces, system state, initialization, and error recovery.

Diagnostic software is often overlooked when considering the embedded software area. This includes the software required to bring up and initialize the hardware, to set up the software task priorities and initial software state, to start the key applications running, and to provide run-time error monitoring, trapping, diagnostic, and repair without upsetting the purpose of the product application.

Issues Confronting Embedded Software Development

The current embedded software methodology is limiting and impedes moving to a platform-based approach for SOC design. It does not promote the reuse and efficient porting of software VCs.

Other major issues and concerns regarding embedding software into SOC designs are as follows:

- What is preventing the emergence of an embedded software industry?
- What are the trends in real-time operating system (RTOS) development?
- How is software ported to a new processor and RTOS?
- How do I simplify generating device drivers?
- What is the current hardware/software co-design and co-verification practice?
- How do I handle verification?

Creating a Software VC Industry

The need for a software VC industry is emerging. All the same factors that contribute to the rise of SOC design (product complexity, sophisticated applications, multidomain applications, increasing time-to-market pressures) apply to the software domain. Software, however, has certain characteristics that make the VC concept difficult to apply as rapidly as desired.[1]

One problem is clear: the rapid proliferation of various hardware platforms in all kinds of embedded application products means that software must be ported to numerous hardware targets. No single processor or architecture dominates, and growth in specific hardware platforms that incorporate processor cores from a host of semiconductor companies means that the number of target platforms continues to proliferate. Since the cost, size, battery life, and other key product factors for embedded portable devices and wired appliances continue to seek differentiation and optimization, the pressures to develop new platforms from a large number of manufacturers continues. Any platform standardization is likely to be deferred and, in fact, might not emerge at all given the continued development of new and varied applications.

Yet all platforms want access to significant amounts of application software content and middleware APIs in order to have rapid product development

---

1. This discussion draws on a presentation made by Sayan Chakraborty, Vice President and General Manager of Development Tools, Cygnus Solutions, entitled "The Riddle of Software IP," made to the VSIA System Level Design Development Working Group (SLD DWG) on March 27, 1998. It relies on general information presented on the software industry only, not on any Cygnus-specific information.

cycles. Hardware differentiation and the desire for standard software development platforms are in opposition. One possible solution is a more rigid architectural approach of dividing the software layers into hardware-dependent and hardware-independent layers. The hardware-dependent layer, which contains hardware and processor and RTOS dependencies, has a well-defined set of interfaces to the layers above it. The hardware-independent layer avoids any specific hardware, processor, and RTOS dependencies and also has well-defined interfaces to layers below and above it. Theoretically, only the hardware-dependent layer requires specific porting to a new hardware platform, and if it is minimized, the effort is less.

Trends in RTOS Development

There are two opposing trends in the evolution of RTOSs. One is the attempt by Microsoft to widen the range of applicability of its Windows CE™ operating system beyond the current applications in portable digital assistants (PDA) to a wider variety of applications that demand "harder" real-time performance.[2] Microsoft is also aiming to apply Windows CE to such areas as embedded automotive use, not in the hard-core engine control and safety-critical areas but in the higher levels of the automotive electronic environment, such as navigation aids and entertainment and mobile office applications.[3]

Microsoft has the financial and technical resources to move Windows CE to a variety of embedded application areas. However, its suitability for hard real-time applications, which also need a very small memory footprint for the RTOS, is open to question. Windows CE is relatively large in extent and contains layers that are suitable for some of the key applications in PDAs and portable entertainment and office appliances (such as the Win32 API), but are overkill in such things as cellular handsets and lower-end embedded appliances. It remains to be seen how future developments in Windows CE will move the market as a whole.

Having Windows CE as a de facto industry standard is a big advantage for application software developers, since porting requirements are greatly reduced, and a large support infrastructure would emerge, such as APIs, other software VCs, and more standardized development systems.

The opposing tendency is for RTOSs to aim for ever-smaller footprints and for significant application specificity to be incorporated. Commercial RTOSs have supported microkernel configuration for a long time, although

---

2. Tom Wong, "The Rise of Windows CE," *Portable Design*, September 1997, pp. 57-58; and Alexander Wolfe, "Windows CE Has a 'Hard' Road Ahead," *Electronic Engineering Times*, April 13, 1998, www.techweb.com/ se/directlink.cgi?EET19980413S00.

3. Terry Costlow, "In-Vehicle PCs Face Bumpy Road Ahead," *Electronic Engineering Times*, March 2, 1998, www.techweb.com/wire/story/TWB19980302S0017.

the degree of parameterization and the ability to include or exclude specific components and features has been relatively coarse. New entrants into the market offer new approaches and claim a much finer-grained approach to creating RTOSs for very specific applications. For instance, Integrated Chipware is a new startup that offers a hierarchical, extensible set of fine-grained component libraries, which can be customized and combined to generate an optimized kernel RTOS. They also claim to have carefully layered the processor-specific portions of the RTOS to minimize the porting effort to a new processor and to layer away the processor-specific dependencies.[4]

Other companies are providing combinations of RTOS and application-specific modules tailored to specific markets. FlashPoint Technology announced a reference design for a wide range of digital cameras based on the WindRiver VxWorks RTOS and specific software modules created by FlashPoint.[5] This was combined with a reference design, creating an integration platform targeted to digital cameras. Some of the customizations involved in this approach include the choice of specific peripheral interfaces. By pre-tailoring specific interfaces in the RTOS and reference design, the company hopes its platform will appeal to a wide variety of end-product customers. In this sense, the platform and the RTOS are inseparably bound together.

Generating a new, application-specific RTOS for an application raises as many questions as it answers. Since the RTOS implementation is new, it is unverified in practical experience and would require extensive validation by software writers and system integrators to achieve the same level of confidence as a well-known commercial RTOS. This argument also holds for switching to a new, commercial, application-specific RTOS that is unfamiliar to a design team. Similarly, either a newly generated RTOS for a specific application or a commercial one targeted to an application space would be quite unfamiliar to many software writers, and the learning curve to use it effectively could be rather steep. In this case, a commercial RTOS would be advantageous, because it has been validated by its creators as well as through substantial commercial use, and also because many software teams are already familiar with it.

Porting Software to a New Processor and RTOS

When porting application software or middleware to a new processor and RTOS, an integration platform could meet the needs of specific derivative

---

4. Terry Costlow, "Startup Lets Developers Custom-design RTOSes," *Electronic Engineering Times*, March 30, 1998, p. 10.

5. Yoshiko Hara and Terry Costlow, "Digital-Camera OS Develops," *Electronic Engineering Times*, April 20, 1998, p. 14.

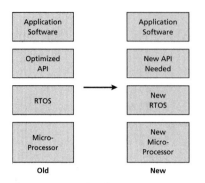

Figure 9.3. Porting from One Processor and RTOS to Another

designs, without requiring significantly more development and porting effort and time, as well as avoiding the added risk of software reimplementation and validation. Another way to reduce porting costs is to abstract common mechanisms, such as tasking and intertask messaging, from specific RTOSs and use this common layer instead of directly using each RTOS's functions. Figure 9.3 illustrates the situation today.

A more appropriate approach would be to have a common RTOS abstraction layer, as shown in Figure 9.4. In this approach, a standard RTOS target API is defined, and the application software has an RTOS-independent layer and an RTOS-aware layer. Because there might be high performance RTOS functions provided under specific RTOSs, a layer of exceptions, which needs to be dealt with manually, should be included. Such high-performance exceptions could contain very application-specific services, which would justify the extra

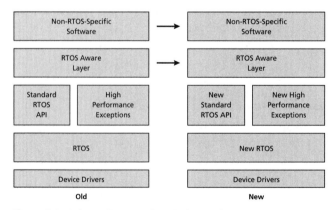

Figure 9.4. Using a Common RTOS Abstraction Layer

porting effort of optimizing the application. However, if a particular software application does not need to deal with such exceptions, the job of porting software to a new RTOS is minimized, requiring merely a relinking to a new API layer that interfaces with the RTOS.

It can be argued that POSIX was an attempt to define a standard RTOS set of functions and services that would allow greater portability of real-time software, and that the POSIX effort failed because it was too basic and offered inadequate performance in comparison to what could be achieved using the special features of commercial RTOSs. In addition, POSIX was criticized for being too arcane and inefficient for real-time embedded systems, since it is based on UNIX.[6] However, ideas that were not viable on a previous generation of technology can often work for the next generation. The evolution of embedded systems may well demand moving to an RTOS standard in the future, at least for certain families or classes of applications.

Simplifying Device Driver Generation

Another area where significant improvements for embedded software are possible is in automated device driver generation, as illustrated in Figure 9.5. This concept, which utilizes the standard Virtual Socket Interface (VSI) Alliance's interface or socket definitions, develops a specification for the interface between the RTOS and a hardware VC function (the device). Development of third-party commercial tools should allow this to be realized over time.

Some commercial tools are now supporting the automated generation of device drivers. For example, Aisys, which is marketing a tool called Driveway 3DE that automates the design of device drivers through tool assistance and templates, claims that their toolset reduces driver development cost by 50

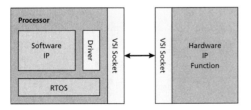

Figure 9.5. Automatic Generation of Device Drivers

6. Jerry Epplin, "Linux as an Embedded Operating System," *Embedded Systems Programming*, October 1997, www.embedded.com/97/fe39710.htm.

percent and time by 70 percent.[7] The tool supports driver development using the following methods:

- Online microcontroller datasheets and interactive help to assist in defining the driver
- Chip browsing tool to identify the peripheral for which the driver is to be generated
- Driver API definition window to define the driver's function and options
- Peripheral configurator to choose specific peripheral options
- Code generation to create the driver code

The automation of these functions will become even simpler as the VC interface standard becomes well defined.

Hardware/Software Co-Design

Current embedded software design involves hardware/software co-design, co-verification, and co-simulation. Co-design starts with functional exploration, goes through architectural mapping, hardware/software partitioning, and hardware and software implementation, and concludes with system integration.

The hardware/software co-design phase divides the system into a set of hardware and software components for which specifications are generated and passed to detailed implementation phases of design.

The activities that occur during the implementation of hardware and software components are referred to as hardware/software co-verification. This includes the verification activities occurring during system integration.

Traditionally, hardware and software implementation processes have often diverged after being based on (hopefully) originally synchronized and compatible specifications from the system design and partitioning stages. To some extent, hardware/software co-verification activities can be regarded as specific attempts to bring these diverging implementation activities back together to validate that the detailed hardware and software designs are still consistent and meet overall system objectives.

Most hardware/software co-verification in system design occurs when the first prototype system integration build occurs, at which point an often lengthy run-debug-modify-run-debug-modify cycle of activity begins.

---

7. Simon Napper, "Tools Are Needed to Shorten the Hardware/Software Integration Process," *Electronic Engineering Times Embedded Systems Special Report*, 1997, available from the Aisys Web site at www.aisys.co.il/news/eetimes.html; and Simon Napper, "Automating Design and Implementation of Device Drivers for Microcontrollers," 1998, available from the Aisys Web site www.aisysinc.com/Community/Aisyp/aisyp.htm.

Table 9.1. Levels of Co-simulation Technology

Abstraction Level	Speed	Debug	Model	Turn-around	Soft-ware	Hard-ware
Nano-second accurate	1-100	Best	Hardest	Fast	OK	Yes
Cycle accurate	50-1000	Excellent	Hard	Fast	OK	Yes
Instruction level	2000-20,000	OK	Medium	Fast	Yes	OK
Synchronized handshake	Limited by hardware sim	No processor state	None	Fast	Yes	OK
Virtual hardware	Fast	No processor or hardware state	None	Fast	Yes	No
Bus functional	Limited by hardware sim	No processor state	Easier	Fast	No	Yes
Hardware modeler	10-50	No processor state	Timing only	Fast	OK	Yes
Emulation	Fast	Limited	None	Slow	OK	OK

Co-simulation technologies vary tremendously in their speed and ability to deal with large testbenches, especially system-level tests. Table 9.1 indicates the relative effectiveness of various co-simulation techniques.[8]

Notwithstanding the range of performance shown in the table, the attempt to integrate views of the hardware and software implementations and to use either commercial or ad hoc methods of hardware/software co-simulation to validate that the system will eventually integrate correctly is doomed to failure unless considerable attention is paid to what needs to be verified at each phase of design and implementation and at what level of abstraction should the verification occur.

Orthogonal Levels of Verification

To address the issue of verification in hardware/software co-design, an orthogonal method can be adopted. The concept of "orthogonalizing concerns" and

---

8. J. Rowson, "Hardware/Software Co-Simulation," *Proceedings of the Design Automation Conference*, 1994, pp. 439–440.

abstracting the questions asked in verifying a design has been discussed in other sources.[9] Essentially, the concept is based on the following:

- Identifying the design's separable levels of abstraction
- Ensuring that appropriate models exist for design components at each level of abstraction
- Identifying the verification questions that can be answered at each abstraction level
- Creating suitable testbenches at each level
- Validating the design at each level, and passing the appropriate testbenches from higher abstraction levels to lower

For most systems, the levels of abstraction shown in Figure 9.6 can be identified.

Of the concepts presented above, solving the verification levels, model access, and technology issues can be done by using some of the emerging commercial or ad hoc hardware/software co-simulation tools. The most challenging issues are identifying what should be validated at each level, constructing the detailed testbench for that level of design, and generating subsidiary or derived testbenches that can be passed from one level to the next. Given the fact that each level of abstraction is between 10 and 1000 times faster in simulation efficiency than the next level down, albeit with a corresponding reduction in fine-grained simulation detail, it behooves the design team to answer each verification question at the highest possible level of abstraction that can reliably furnish an appropriate answer.

For example, when a cellular phone call in a moving vehicle is handed off from one cell to the next, moving from one base station's control to the next, a

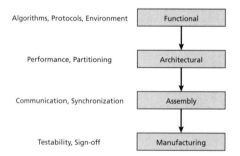

Figure 9.6. Levels of Design Abstraction

9. Alberto L. Sangiovanni-Vincentelli, Patrick C. McGeer, and Alexander Saldanha, "Verification of Electronic Systems," *Proceedings of the Design Automation Conference*, 1996, pp. 106-111; J. Rowson and A. Sangiovanni-Vincentelli, "Interface-based design," *Proceedings of the 34th Design Automation Conference*, 1997, pp. 178-183; and C. Ussery and S. Curry, "Verification of large systems in silicon," *CHDL 1997*, Toledo, Spain, April 1997.

number of interesting control events occur while speech is being processed to ensure smooth handoff. High-level system design tools for dataflow systems can be used to verify the correct handling of speech in the baseband algorithmic processing layer, and the correct interactions between baseband and protocol layers during the control processing. A blind mapping of this testbench from the algorithmic level of abstraction to a register-transfer level (RTL) and C-level of abstraction, using commercial hardware/software co-simulation techniques, results in a testbench that runs for an impracticably enormous number of clock cycles, since the level of abstraction is much lower. The most interesting fact about such a testbench is that a huge part of it, and a very large part of the resulting simulation time, is validating the same thing over and over: the correct handling of speech by the hardware/software combination involved in baseband processing. Only a very small part of the simulation is dealing with the control events that occur and the interactions between protocol and baseband processing.

It is therefore important to subset or segment higher-level system testbenches into slices that deal only with the most important verification questions at the lower level, such as hardware/software interactions and control processes that deal directly with the handoff, and task scheduling and interrupt handling of baseband processing on the DSP in the handset, rather than the mundane and well-proven baseband processing algorithms. This concept is illustrated in Figure 9.7. Methodologies and tools to allow accurate and complete testbench segmentation for lower-level verification are currently at a very primitive state of development, and robust techniques that solve this problem will likely take some time to develop.

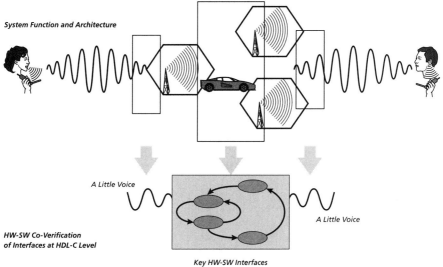

Figure 9.7. Orthogonal Verification by Slices

Improving Embedded Software Development Methodology

To improve the current development methodology for embedded software so that development time of derivative products can be shortened and risk reduced, the following methods need to be adopted:

- The hardware/software definition, trade-offs, partitioning, and modeling are done at the system level.
- The software architecture is an integral part of the application-oriented integration platform.
- The software architecture is carefully layered to minimize porting to new processors and RTOSs.
- Application-specific RTOSs are defined as part of the platform, when appropriate.
- Software is structured to maximize reusing software VCs.
- Further develop standards, such as the VC Interface, to help automate device driver development.

If we return to the questions asked earlier in the chapter, we can propose the following solutions and approaches.

Table 9.2. Solutions to Embedded Software Issues

Issue	Recommended Solution
What is preventing the emergence of an embedded software industry?	Rapid proliferation of hardware platforms (various embedded processors) and a profusion of RTOSs means that most tentative software VCs must be tediously ported and reoptimized for each processor-RTOS combination. This is especially true with assembly code, for example, in various DSP applications.
What are the trends in RTOS development?	Homogenization, as might emerge with new versions of Microsoft Windows CE, which are hard real-time. Differentiation—variety of RTOSs, new microkernels, new ideas for automated configuration and generation.
How is software ported to a new processor and RTOS?	To be efficient, software needs better APIs and layering to isolate processor and RTOS dependencies. Standardized APIs for device drivers, scheduling, and messaging services also need to be offered. Currently, this still involves extensive manual development and testing efforts.
How do I simplify generating device drivers?	There are new approaches to template-driven device driver synthesis for at least some classes of processors and devices.
What is the current hardware/software co-design and co-verification practice?	Very ad-hoc. Some emerging co-simulation tools and practices, but little standardization. Often little understanding of which techniques are available and how to use them.
How do I handle verification?	The method of orthogonalizing verification concerns. Cut the verification problem into layers and use various abstraction levels and models, and co-simulation techniques at different levels. Also reduce test cases as one goes down the verification stack by using slices of the upper test cases or targeted special-purpose tests.

10

In Conclusion

This book has explored the challenges and benefits of transitioning to SOC design, and the methodology changes required to meet them. These methodology changes include function-architecture co-design, bus-based communications architectures, integration platforms, analog/mixed signal (AMS) integration, and embedded software development. Each of these methodologies fits into and leverages platform-based design, which we believe is the only method that is going to meet the design productivity challenge. But it is when all these approaches are brought together that platform-based IC design is the most productive.

Economics—the Motivator

The primary driver for the methodology shifts we have been discussing is the same as what motivates most business changes—economics. However, the economics that are driving the SOC transition are unique in terms of how significant the shifts are and how quickly they are happening. The economics of the semiconductor industry, which are now taken as law, are that each successive generation of IC technology enables the creation of products that are significantly faster, consume much less power, and offer more capabilities in a smaller form factor, all at a greatly reduced system cost. These changes drive, as well as are driven by, the growing consumer desire for electronics-fortified products. The combination of a new generation of IC technology occurring every 18 months (and accelerating) and the continuing growth of worldwide consumer markets has become a considerable, if not the largest, economic lever in the world.

Effectively leveraging the capabilities of the semiconductor process technology into short-lived, rapidly evolving consumer products is a high-stakes game with big win or big loss potential. We have seen industry leaders in one product generation be completely overwhelmed by the next generation. For example, in the case of modems, Hayes was replaced by Rockwell, who was replaced by Texas Instruments and Lucent. The winners in this race were able to respond to the market demands by capitalizing on the latest generation of process technology and creating a product that was differentiated by new features. While the ability to see the emerging market and to have the vision of how to realize it using new process technology lies at the heart of most successful companies, by not adopting a strong design methodology, a company could still fail at producing a timely product.

Platform-Based Design—the Enabler

Platform-based design is an essential element in producing comprehensive SOC designs. As discussed in Chapter 3, different levels of platforms address the trade-offs between flexibility and productivity. The expectation is that as the IC industry pursues the process technology evolution, companies will follow this methodology evolution. However, different application markets will follow the evolution at different rates, based upon their particular evaluation of economic factors and their ability to assimilate the methodology changes.

Platform levels 2 and above have the potential to fulfill the Virtual Socket Interface Alliance's goal of plug and play SOC design. The plug and play goal establishes that all the connections between the block and the integration platform are completely specified (the plug), as well as the operational behavior, the data, and the instruction streams (the play). Thus, the integration effort for SOC design can be essentially reduced to only the physical design merge and layout verification. This level of design productivity provides an answer to the "design productivity gap" between cost-effectively designing systems on-chip and manufacturing them. SOC design productivity would then become more analogous to printed circuit board (PCB) design, where the design and verification focus is primarily on the interactions between components on the board, and not on the design and verification of the components themselves. This move upward in the abstraction level for IC design is fundamentally enabled by the platform.

Platform-based design also provides the opportunity to better leverage emerging design technologies and methodologies. As we have discussed throughout this book, the move to platform-based design results in productivity gains through partitioning of the design problem to better support function-architecture co-design, separation of core functions and interface design, the effective integration of analog circuits into digital SOCs, and a modular embedded software architecture that parallels the hardware architecture. These combined benefits yield a solution path that is sure to evolve beyond what we have outlined.

SOC—Changing How We Do Things

In this book, we have attempted to cover the broad spectrum of SOC design. However, practicalities and conceptual maturity dictated that some topics were not fully explored. We would like to mention some areas that merit more attention as SOC methodology develops.

For instance, an entire book could be devoted to the manufacturing test issue. The methods for testing systems and component chips are traditionally quite different. When systems are implemented on boards, testing is amenable to a naturally hierarchical approach: test the components using a variety of techniques, mostly scan- or BIST-based; test the components in-circuit on the board after assembly; if a component is bad, remove and replace it; if the board is bad, repair it. Tests for the board are largely connectivity-based, followed by running the system itself. This process does not translate well to SOC applications. As systems are implemented on SOCs, a serious crisis in the testing approach occurs.

The method outlined in Chapter 7 is characteristic of approaches taken. However, the time required for a manufacturing tester to perform detailed stuck-at-fault, embedded memory, at-speed performance, and AMS tests for potentially hundreds of blocks is excessive by today's IC standards. SOC designs will require new trade-offs in the areas of coverage, failure isolation, cost, and reliability. The use of platform-based design to create solutions to these trade-offs will emerge and mature. Using on-chip processors to perform the tests will evolve over time to a unique test architecture potentially populated with dedicated processors, reconfigurable design elements, redundancy options, and additional interconnect overhead to ensure system economics and constraints are satisfied in manufacturing. However, even assuming significant advances in tester technology, the final solution will require a recognition and acceptance that the economics of testing SOC designs needs to be adjusted to reflect the intellectual content of the device under test and the reliability requirements of the end user.

Microelectronic mechanical systems (MEMS) will also demand a significant rethinking in regards to SOC design. The foundation established for predetermining at the platform-design level the relationship between analog and digital logic, which extends down to the substrate design level, can serve as a starting point.

Another area to be examined is chip package-board design. As systems are put on chips, the packaging and board interface issues associated with high-speed, real-world interfaces will demand that custom package design become the norm rather than the exception. Issues, such as simultaneous switching, heat, multi-power design, noise, and interfaces to off-chip buses, will need to be analyzed at a level consistent to what is being done at high-performance system houses today.

A Change in Roles

All these topics, as well as extensions to the embedded software reuse model, the utilization of reconfigurable hardware, and the emergence of a viable intra-company VC business model, are subjects for future exploration. However, the real challenge of the future lies in our ability to adapt to what is the most rapidly expanding set of technology-driven opportunities in the past two decades.

The transition to SOC is realigning the entire industry and affecting who is contributing to IC design. This significant change, which is taking many years to roll out, is contributing to an overall *deverticalization* of the IC industry. A single company is no longer exclusively handling the design for a specific IC, instead many companies are contributing based upon their specific areas of expertise. This leads to a group of companies linked together to provide the complete IC design solution. This type of change is causing companies to re-engineer themselves and determine where their core competencies are—and where they aren't.

No longer do design engineers have neatly demarcated descriptions separating architects from logic designers from IC designers from software designers. The skill sets required to design a product or author a reusable VC demand a breadth that harkens back to the "tall, thin designer" analogies of the early 80s. The opportunities for the individual engineer to expand his or her horizons, to move beyond the bounded roles defined in the ASIC model, are greater now than ever before. But it will require an unprecedented cooperation among teams, across groups, within and across industries.

This reorganization, investment, and divestment, added to a methodology change, presents new challenges and opportunities. Those companies who move the fastest and see the new opportunities can seize them, and perhaps displace the market leaders who are clinging to the previous methodology and industry roles.

The Golden Age of Electronics

With a cost-effective means of designing, verifying, and building complex SOCs, labeling this generation the Golden Age of Electronics will be hard to dispute. The semiconductor industry will affect every electronic device as well as many non-electronic devices that can benefit by adding some form of electronic intelligence or interconnection. Imagine a $20 add-on to an existing "dumb" device that can anticipate its own and your needs and respond to them automatically, learn usage profiles, use resources more efficiently, be safer, and connect with the Internet and other devices. This add-on is well within the range of the SOC designs that we have been discussing in this book. An embedded microcontroller, sensor inputs and actuator outputs (analog to digital, digital to analog), network interfaces (wireless, wireline), and some embedded software provide the necessary functionality. The single-chip implementation offers the cost point that creates the economic impetus to deliver these add-ons.

Previously, either the functionality or the cost (or both) was out of reach for most products. With the latest process technologies and efficient SOC design (platform-based), many applications instantly become possible. Adding electronics intelligence to non-typical electronics devices (electronics infusion) can bring new life to stagnant products.

The impact of SOC goes beyond the microwave oven that automatically scans the bar code on the food package and then looks on the Web for the appropriate cooking time, which is scaled to the power of the oven, ambient temperature, and estimated cooking completion time of the rest of the meal on the "smart" stove. SOC also provides the low cost, low power, high performance, small form factor devices that will fulfill the prophesy of the disaggregated computer and ubiquitous computing. The combination of these two major changes, electronics infusion and ubiquitous computing, leads to a world of intelligent, interconnected everything. The efficient application of SOC technology through platform-based design could lead to continued and lasting productivity gains across the economy. Then, we electronics designers can truly say we changed the world!

Index